建筑立场系列丛书 No.42

木建筑再生
Recovering Wood

中文版
(韩语版第358期)
韩国C3出版公社 编
杨惠馨 王平 李海玲 赵伟丽 王莹 周一 时跃 郭薇 译

大连理工大学出版社

004 坂茂荣获2014年普利兹克建筑奖
006 2014年蛇形画廊展馆 _ Smiljan Radic
010 阿尔罕布拉场址的新入口和游客中心
　　 _ Álvaro Siza Viera + Juan Domingo Santos
014 高塔vs天线 _ Smiljan Radic + Gabriela Medrano + Ricardo Serpell
018 电子谷 _ OMA
022 本地建筑的普适性 _ YongJu Lee
026 柳条剧院 _ Tim Lai Architect + Brad Steinmetz Stage Design
030 Kineforum Misbar _ Csutoras & Liando

建筑脉动
形式的张力
036 形式的张力 _ Silvio Carta
040 爱默生学院洛杉矶中心 _ Morphosis Architects
054 奥尔堡音乐之家 _ Coop Himmelb(l)au

起居与工作相融合
064 起居与工作相融合 _ Douglas Murphy
068 SISII花园式展示厅 _ Yuko Nagayama & Associates
074 Koen Van Den Broek工作室 _ Haerynck Vanmeirhaeghe Architecten
084 Draft工作室与住宅 _ Estudio Elgue
096 绿色办公室 _ Fraher Architects
102 地下车库建筑工作室 _ Carlo Bagliani
110 稻田中的办公室 _ Akitoshi Ukai / AUAU
118 Roduit工作室 _ Savioz Fabrizzi Architectes

木建筑再生
124 木建筑再生 _ Tom Van Malderen
128 Tamedia办公大楼 _ Shigeru Ban Architects
138 主教爱德华国王礼拜堂 _ Níall McLaughlin Architects
150 杰克逊霍尔机场 _ Gensler
158 皮塞克市林业局 _ HAMR
170 园艺展览中的展亭 _ Dethier Architecture
182 Reussdelta瞭望塔 _ Gion A. Caminada

188 建筑师索引

004 Shigeru Ban receives the 2014 Pritzker Architecture Prize
006 Serpentine Gallery Pavilion 2014 _ Smiljan Radic
010 New Entrance and Visitor Center of Alhambra
     _ Álvaro Siza Viera + Juan Domingo Santos
014 Tower vs Antenna _ Smiljan Radic + Gabriela Medrano + Ricardo Serpell
018 Digital Valley _ OMA
022 Vernacular Versatility _ YongJu Lee
026 Willow Theater _ Tim Lai Architect + Brad Steinmetz Stage Design
030 Kineforum Misbar _ Csutoras & Liando

Archipulse
A Formal Tension

036 *A Formal Tension _ Silvio Carta*
040 Emerson College Los Angeles Center _ Morphosis Architects
054 House of Music in Aalborg _ Coop Himmelb(l)au

Live/Work Hybrids

064 *Live/Work Hybrids _ Douglas Murphy*
068 SISII _ Yuko Nagayama & Associates
074 Koen Van Den Broek Studio _ Haerynck Vanmeirhaeghe Architecten
084 Draft Studio and House _ Estudio Elgue
096 The Green Studio _ Fraher Architects
102 Underground Garage Architecture Office _ Carlo Bagliani
110 Office in Rice Field _ Akitoshi Ukai / AUAU
118 Roduit Studio _ Savioz Fabrizzi Architectes

Recovering Wood

124 *Recovering Wood _ Tom Van Malderen*
128 Tamedia Office Building _ Shigeru Ban Architects
138 Bishop Edward King Chapel _ Níall McLaughlin Architects
150 Jackson Hole Airport _ Gensler
158 Pisek City Forest Administration _ HAMR
170 Pavilion for Horticultural Show _ Dethier Architecture
182 Reussdelta Observation Tower _ Gion A. Caminada

188 Index

坂茂荣获2014年普利兹克建筑奖
Shigeru Ban receives the 2014 Pritzker Architecture Prize

2014普利兹克建筑奖授予了日本的建筑师坂茂。

建筑师坂茂生于日本东京，现年56岁，现在在东京、巴黎和纽约设有工作室，是建筑界的奇才。他将同样的、具有创造性的、丰富的设计方法广泛应用于人道主义事业。20年来，坂茂一直奔波于自然及人为灾难现场，与当地居民、志愿者和学生合作，为灾民设计建造简单、体面、低成本并且可循环使用的避难场所和社区建筑。

坂茂在自己位于巴黎的办公室里接受采访时表示："接受这个奖项对我来说是一个莫大的荣誉。我必须更加认真谨慎。我必须继续聆听于我服务的对象，无论他们是私人住宅客户还是赈灾工作中的人们。我将这个奖项视为对自己的鼓励，继续坚持自己的事业——不会有所改变，只会继续成长。"

坂茂在他从事的所有领域总是能够发现多种多样的设计方案，他通常会根据结构、材料、景观、自然通风和光照条件，致力为建筑物的使用者们营造舒适环境。从私人住宅、企业总部，到博物馆、音乐厅和其他民用建筑，坂茂的作品总是以其创造性、经济性和精巧性著称，并不依赖现代普遍的高科技方法。

为了建造赈灾避难所，坂茂常常采用可循环使用的硬纸管作为柱子、墙壁和房梁，因为它们容易取材；价格低廉；方便运输，安装和拆卸；并且防水、防火、可循环利用。他说在日本的成长经历培养了自己不浪费任何材料的信念。

孩提时代，坂茂曾观察传统日本木匠在他父母的住宅内工作的情形。在他的眼里，他们的工具、施工过程，甚至是木材的气味都那么神奇。他会收集废弃的小块木材，用它们来搭建小模型。他的理想是成为一名木匠。但在坂茂11岁那年，老师让全班同学设计简单的房屋，坂茂的作品被评为最佳并在全校展览。从那时起，他就一直梦想成为一名建筑师。

坂茂的人道主义事业始于1994年的卢旺达大屠杀，数百万人流离失所。坂茂向联合国难民事务所的高级专员提出了用硬纸管建造避难所的想法，并被聘为顾问。1995年日本神户大地震以后，他再次贡献出了自己的时间和才能。在那里，坂茂为当地的越南难民开发出了"纸木宅"，利用人们捐赠的啤酒箱装满沙袋作为地基，将硬纸管垂直排列形成房屋的墙壁。坂茂还设计了"纸教堂"，作为神户地震受灾民众的社区活动中心。它后来被拆散，并于2008年被运往台湾重建。

评委会还特别提到了他于2000年在日本埼玉县设计的"裸宅"，坂茂用透明的瓦楞塑料板覆盖外墙，又在木质框架上绷白色亚克力形成室内墙面。透明板隔层令人联想起了泛光的日式障子。客户要求不能有家庭成员被孤立，所以房子只是一个独特的大型空间，有两层楼那么高，四间个人居室都安装了脚轮，能够自由移动。

在东京的"幕墙住宅"（1995年）中，沿房子外围的两层高的白色窗帘可以打开，进行空气对流，也可以关闭形成一个茧状的背景空间。

东京14层高的尼古拉斯·G. 海耶克中心（2007年），以其前后外立面均安装了高大的玻璃百叶窗为特点，而且窗子可以完全打开。

坂茂曾使用运输用的集装箱作为现成的原材料，建造了游牧博物馆（纽约，2005年；圣莫尼卡，加利福尼亚，2006年；东京，2007年）。

他设计的阿斯彭艺术博物馆于2014年8月开幕。

坂茂曾在2006年至2009年间担任普利兹克建筑奖的评委会成员。他还在世界各地多所建筑院校讲学和任教，目前是京都造型艺术大学的教授。

坂茂最初在南加利福尼亚建筑学院（当时的总部设在圣莫妮卡，加利福尼亚）接受教育，并于1984年在纽约库伯联盟学院获得了建筑学学士学位。

坂茂将成为第七位荣获普利兹克建筑奖的日本建筑师——前六位分别是已故的丹下健三（1987年）、桢文彦（1993年）、安藤忠雄（1995年）、妹岛和世与西泽立卫建筑团队（2010年），以及伊东丰雄（2013年）。

颁奖仪式于2014年6月13日在荷兰阿姆斯特丹国立博物馆举行。每年的普利兹克建筑奖颁奖仪式都会在世界上具有重大文化或历史意义的场所举办。今年将首次在荷兰举行。

海地的纸质应急避难所，太子港，海地，2010年
Paper Emergency Shelter for Haiti, Port-au-Prince, Haiti, 2010

纸木宅，布吉市，印度，2001年
Paper Log House, Bhuj, India, 2001

临时集装箱住宅，神奈川，宫城县，日本，2011年
Container Temporary Housing, Onagawa, Miyagi, Japan, 2011

纸木宅，神户，日本，1995年
Paper Log House, Kobe, Japan, 1995

克林达住宅，克林达，斯里兰卡，2007年
Kirinda House, Kirinda, Sri Lanka, 2007

纸质隔间系统4，日本，2011年
Paper Partition System 4, Japan, 2011

Shigeru Ban received the 2014 Pritzker Architecture Prize.

A Tokyo-born, 56-year-old architect with offices in Tokyo, Paris and New York, is rare in the field of architecture. He uses the same inventive and resourceful design approach for his extensive humanitarian efforts. For twenty years Ban has traveled to sites of natural and man-made disasters around the world, to work with local citizens, volunteers and students, to design and construct simple, dignified, low-cost, recyclable shelters and community buildings for the disaster victims.

Reached at his Paris office, Shigeru Ban said, *"Receiving this prize is a great honor, and with it, I must be careful. I must continue to listen to the people I work for, in my private residential commissions and in my disaster relief work. I see this prize as encouragement for me to keep doing what I am doing – not to change what I am doing, but to grow."*

In all parts of his practice, Ban finds a wide variety of design solutions, often based around structure, material, view, natural ventilation and light, and a drive to make comfortable places for the people who use them. From private residences and corporate headquarters, to museums, concert halls and other civic buildings, Ban is known for the originality, economy, and ingeniousness of his works, which do not rely on today's common high-tech solutions.

To construct his disaster-relief shelters, Ban often employs recyclable cardboard paper tubes for columns, walls and beams, as they are locally available; inexpensive; easy to transport, mount and dismantle; and they can be water- and fire-proofed, and recycled. He says that his Japanese upbringing helps account for his wish to waste no materials.

As a boy, Shigeru Ban observed traditional Japanese carpenters working at his parents' house and to him their tools, the construction, and the smells of wood were magic. He would save cast-aside pieces of wood and build small models with them. He wanted to become a carpenter. But at eleven, his teacher asked the class to design a simple house and Ban's was displayed in the school as the best. Since then, to be an architect was his dream.

Ban's humanitarian work began in response to the 1994 conflict in Rwanda, which threw millions of people into tragic living conditions. Ban proposed paper-tube shelters to the United Nations High Commissioner for Refugees and they hired him as a consultant. After the 1995 earthquake in Kobe, Japan, he again donated his time and talent. There, Ban developed the "Paper Log House," for Vietnamese refugees in the area, with donated beer crates filled with sandbags for the foundation, he lined up the paper cardboard tubes vertically, to create the walls of the houses. Ban also designed "Paper Church," as a community center of paper tubes for the victims of Kobe. It was later disassembled and sent to Taiwan, and reconstructed there, in 2008.

The jury cites Naked House (2000) in Saitama, Japan, in which Ban clads the external walls in clear corrugated plastic and sections of white acrylic stretched internally across a timber frame. The layering of translucent panels evokes the glowing light of shoji screens. The client asked for no family member to be secluded, so the house consists of one unique large space, two-story high, in which four personal rooms on casters can be moved about freely.

In Curtain Wall House (1995) in Tokyo, two-story-high white curtains along the perimeter of the house can be opened to let the outside flow in or closed to provide a cocoon-like setting. The 14-story Nicolas G. Hayek Center (2007) in Tokyo features tall glass shutters on the front and back facades that can be fully opened.

Ban used transportation containers as ready-made elements to construct the Nomadic Museum (New York, 2005; Santa Monica, California, 2006; Tokyo, 2007). His design for the Aspen Art Museum is slated to open in August 2014.

Shigeru Ban served as a member of the Pritzker Architecture Prize jury from 2006 to 2009. He lectures and teaches at architecture schools around the world and is currently a professor at Kyoto University of Art and Design. Ban attended architecture school first at the Southern California Institute of Architecture (then based in Santa Monica, California), and earned his bachelor's degree in architecture from Cooper Union in New York City in 1984.

Shigeru Ban will be the seventh Japanese architect to become a Pritzker Laureate – the first six being the late Kenzo Tange in 1987, Fumihiko Maki in 1993, Tadao Ando in 1995, the team of Kazuyo Sejima and Ryue Nishizawa in 2010, and Toyo Ito in 2013. The award ceremony takes place on June 13, 2014, at the Rijksmuseum in Amsterdam, the Netherlands. The Pritzker Prize ceremony is held each year at a culturally or historically significant venue around the world. This marks the first time the ceremony will be held in the Netherlands.

纸教堂, 神户, 日本, 1995年
Paper Church, Kobe, Japan, 1995

梅斯蓬皮杜中心, 法国, 2010年
Centre Pompidou Metz, France, 2010

纸板大教堂, 基督教堂, 新西兰, 2013年
Cardboard Cathedral, Christchurch, New Zealand, 2013

纸音乐厅, 拉奎拉, 意大利, 2011年
Paper Concert Hall, L'Aquila, Italy, 2011

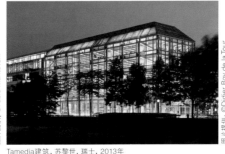
Tamedia建筑, 苏黎世, 瑞士, 2013年
Tamedia Building, Zurich, Switzerland, 2013

纸板大教堂, 基督教堂, 新西兰, 2013年
Cardboard Cathedral, Christchurch, New Zealand, 2013

2014年蛇形画廊展馆_Smiljan Radic

蛇形画廊委任智利建筑师Smiljan Radic来设计2014年蛇形画廊展馆。Smiljan Radic是第十四任接受邀请设计临时蛇形画廊展馆的建筑师,该建筑位于肯星顿花园蛇形画廊的入口外侧。展馆的委托设计是每年文化节日程上最受瞩目的大事之一,而展馆自2000年以来成为伦敦夏季的首席景点。Smiljan Radic的设计沿袭了藤本壮介的云形状结构,因为这个结构在2013年吸引了200 000人前来参观,是迄今为止参观人数最多的画廊展馆之一。

展馆位于蛇形画廊的草地之上,占地350m²,其平面勾勒出一个半透明的圆柱状结构,形似一个在大型采石场静止的贝壳。Smiljan Radic的设计灵感源自他的早期作品,尤其是受奥斯卡·王尔德的启发而设计出的自私巨人城堡项目和部分依傍巨石建成的Mestizo餐厅。展馆设计成一处灵活的多功能空间,内设一座咖啡馆,以在四个月的有效期内吸引游客进入画廊并以不同的方式和展馆进行互动。在7月至9月的每个周五的晚上,展馆将成为由COS赞助举办的蛇形公园晚会系列的舞台,包括8个特定场地的活动。届时会有艺术、诗歌、音乐、电影、文学和理论专场,同时还会发布三个初出茅庐的建筑师莉娜·莱格莱特、汉娜·佩里和希瑟·菲力森的委任书,2014蛇形画廊展馆将在2014年伦敦建筑节期间正式对外开放。

第十四届蛇形画廊展馆的设计者Smiljan Radic称,"2014蛇形画廊展馆是位于公园或大型花园,或所谓的大厅中的小型浪漫主义型建筑的一部分,自16世纪末开始至19世纪初深受欢迎。此外,游客从外边只会看到一个易碎的贝壳悬在巨石上,但是贝壳呈现如下特点:洁白、透明、玻璃纤维质地;内景围绕一空置的露台布置,自然场景在低处可见,营造出整体体量的悬浮感。夜晚来临,半透明的贝壳和琥珀色的光线会吸引路人前来一探究竟,好似飞蛾难敌灯光的诱惑一般。"

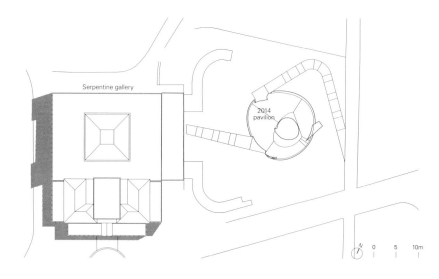

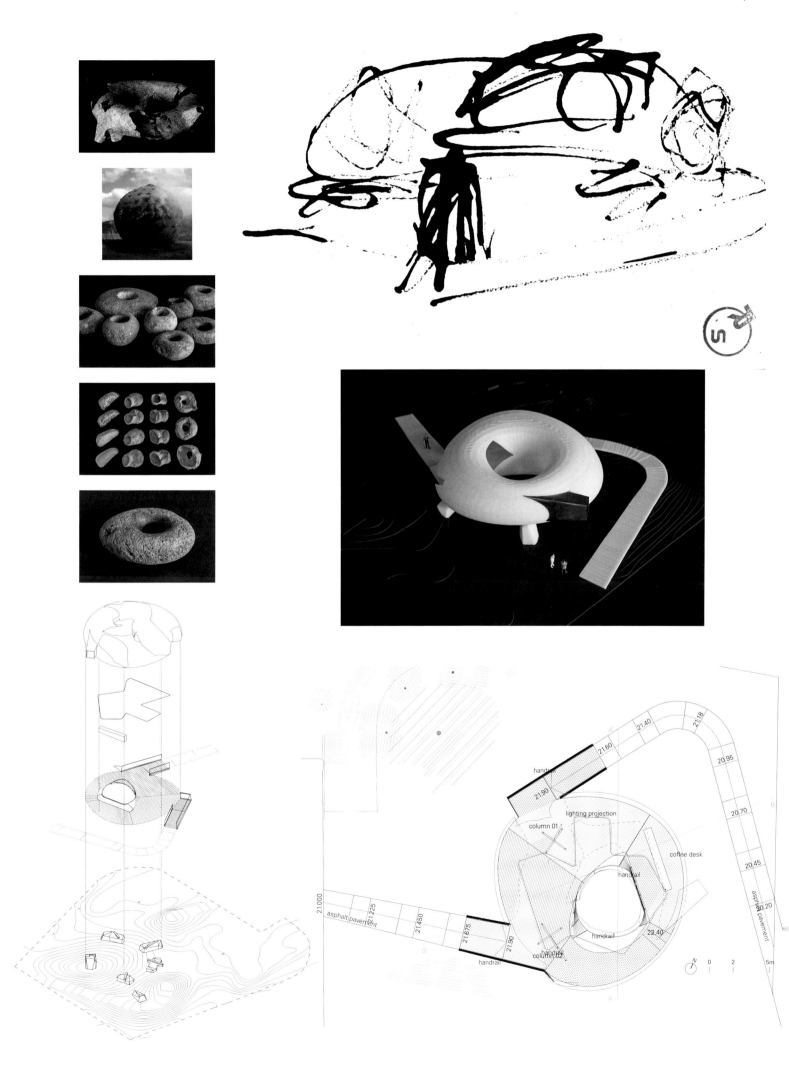

Serpentine Gallery Pavilion 2014

The Serpentine has commissioned Chilean architect Smiljan Radic to design the Serpentine Gallery Pavilion 2014. Radic is the fourteenth architect to accept the invitation to design a temporary pavilion outside the entrance to the Serpentine Gallery in Kensington Gardens. The commission is one of the most anticipated events in the cultural calendar, and has become one of London's leading summer attractions since launched in 2000. Smiljan Radic's design follows Sou Fujimoto's cloud-like structure, which was visited by almost 200,000 people in 2013 and was one of the most visited Pavilions to date.

Occupying a footprint of some 350 square metres on the lawn of the Serpentine Gallery, plans depict a semi-translucent, cylindrical structure, designed to resemble a shell, resting on large quarry stones. Radic's Pavilion has its roots in his earlier work, particularly The Castle of the Selfish Giant, inspired by Oscar Wilde's story, and the Restaurant Mestizo, part of which is supported by large boulders. Designed as a flexible, multi-purpose social space with a cafe sited inside, the Pavilion will entice visitors to enter and interact with it in different ways throughout its four-month tenure in the Park. On selected Friday nights, between July and September, the Pavilion will become the stage for the Serpentine's Park Nights series, sponsored by COS: eight site-specific events bring together art, poetry, music, film, literature and theory and include three new commissions by emerging artists Lina Lapelyte, Hannah Perry and Heather Phillipson. Serpentine Gallery Pavilion 2014 launches during the London Festival of Architecture 2014.

Smiljan Radic, designer of the fourteenth Serpentine Pavilion, said:

"The Serpentine 2014 Pavilion is part of the history of small romantic constructions seen in parks or large gardens, the so-called follies, which were hugely popular from the end of the 16th Century to the start of the 19th. Externally, the visitor will see a fragile shell suspended on large quarry stones. This shell – white, translucent and made of fibreglass – will house an interior organised around an empty patio, from which the natural setting will appear lower, giving the sensation that the entire volume is floating. At night, thanks to the semi-transparency of the shell, the amber tinted light will attract the attention of passers-by, like lamps attracting moths."

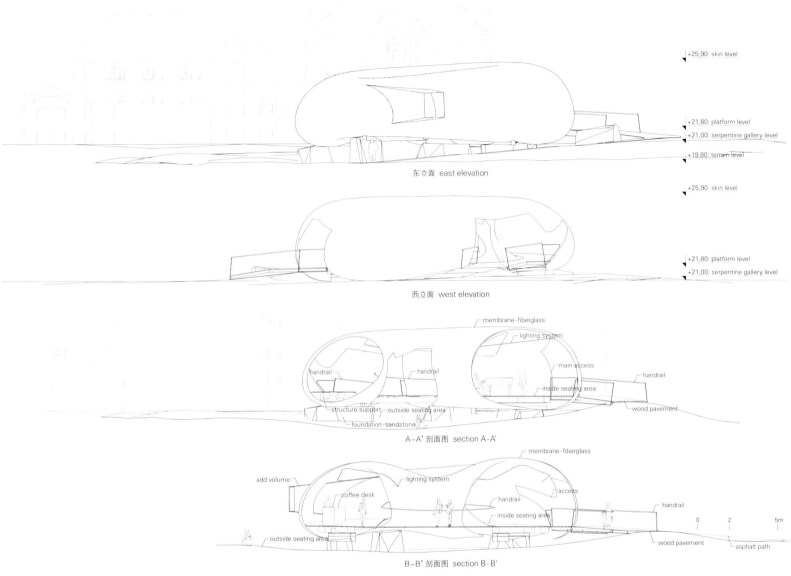

阿尔罕布拉场址的新入口和游客中心 _Alvaro Siza Viera + Juan Domingo Santos

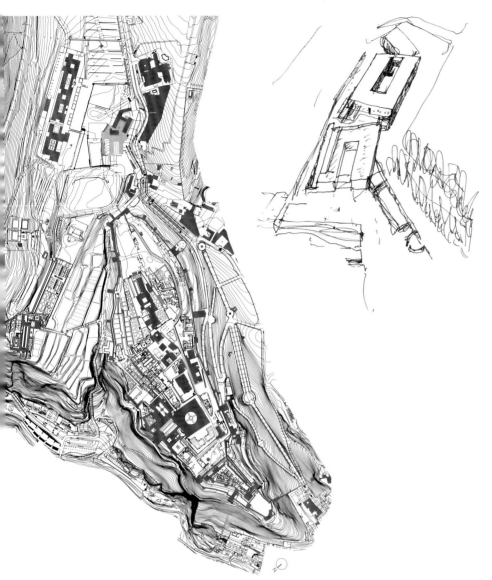

国际设计大赛得主Alvaro Siza Viera和Juan Domingo Santos最近公布了在柏林展览的阿尔罕布拉世界遗产场址的新入口和游客中心的设计方案。

新大门重新布局了进入路线，利用一系列封闭且遮阴的院子以及开阔且充满光线的平台来引导游客，来进入这个历时1000多年的纪念碑。仿照2009年参观此地时留下的感受，Siza将他的视野融入新大门的改建中并称，"从阳光下进入阴影里，从温暖中步入阴凉中，从开阔的场地走入熟悉的焦点区，在蓝图变成细节前我喜欢先在头脑中幻想我的设计。"

建筑师计划用半隐半现的入口大门来代替先前的双层队列体系，游客可以循着光和影交替的地方进入院子，然后到达主广场和礼堂。

展览于3月22日在柏林的Aedes am Pfefferberg画廊举行。主要展出原始素描和模型，同时展示记录与Siza一生为之倾心迷恋的古代宫殿相关的采访和绘画，而这座宫殿现已入选联合国教科文组织的世界遗址。

继柏林之后，展览将移址于莱茵河畔魏尔的维特拉设计博物馆，时间从6月13日至8月31日，下一场将于2015年春季在阿尔罕布拉综合大楼展出。

New Entrance and Visitor Center of Alhambra

Alvaro Siza Viera and Juan Domingo Santos, the winners of international design competition recently unveiled their de-

sign for a new entrance and visitor center of the Alhambra World Heritage site at the exhibition in Berlin.

The new gate rearranges visitor access into the more than 1000-year-old monument through a series of enclosed, shaded courtyards and open, sunlit terraces. Following to his experience at the Alhambra in 2009, Siza journaled about his envision for the new gate, stating:

"from bright sun to shadows, from warmth to coolness, from wide to intimate focus, I like to dream about my project before I set it down in any detail."

The architects plan to replace the existing two-stage queue system, creating a new partially submerged entrance gate that will lead visitors through areas of light and shade towards a courtyard, before arriving at a main plaza and auditorium.

The exhibition was open at the Aedes am Pfefferberg Gallery in Berlin on March 22. It features original sketches and models, plus interviews and drawings documenting Siza's lifelong fascination with the ancient palace which is named as a UNESCO World Heritage Site.

After Berlin, the exhibition will be presented at the Vitra Design Museum in Weil am Rhein between 13 June to 31 August, before moving to the Alhambra Complex for a show in the spring of 2015.

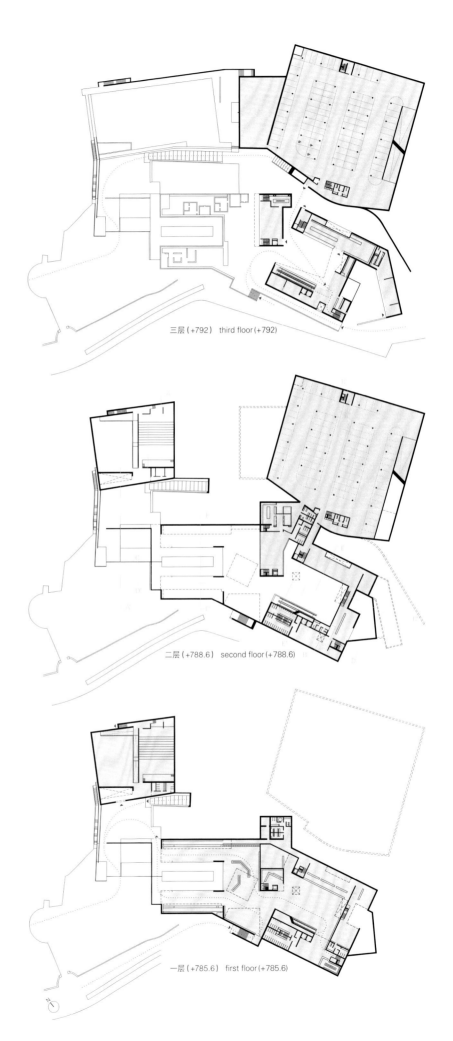

三层(+792)　third floor (+792)

二层(+788.6)　second floor (+788.6)

一层(+785.6)　first floor (+785.6)

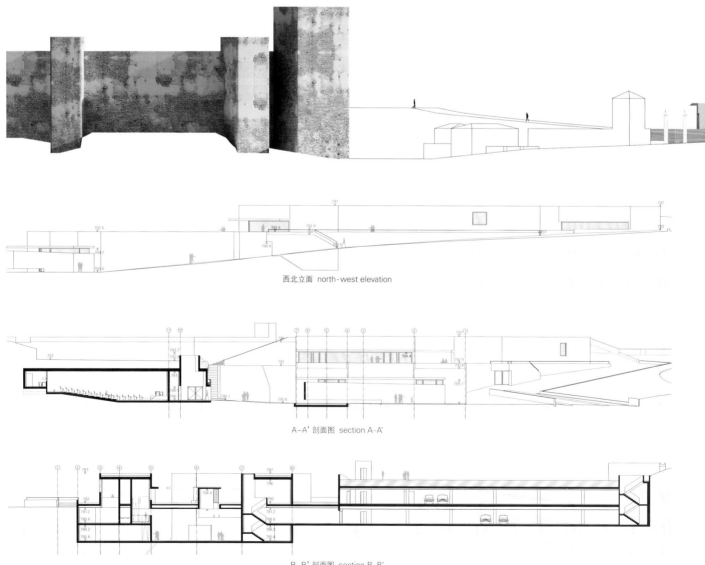

西北立面 north-west elevation

A-A' 剖面图 section A-A'

B-B' 剖面图 section B-B'

项目名称：Alhambra Atrium, Granada, Spain
地点：Granada, Andalusia, Spain
建筑师：Alvaro Siza, Juan Domingo Santos
合作建筑师：Avelino Silva, Hans Ola Boman, Daniel Gutierrez Peinado, Jose Pedro Silva, Ina Valkanova, Lucia Balboa Quesada, Carmen Moreno Alvarez, Isabel Diaz Rodriguez, Julien Fajardo, Claire de Nutte, Carlos Gor Gomez, Pablo Fernandez Carpintero
工程师：Jorge Nunes da Silva, Raquel Fernandez, Alexandre Martins, Alvaro Raimundo, Raul Bessa
工业工程师：ABACO Ingenieros, Patricio Bautista Carrascosa
农业工程师：Rafael M. Navarro Serrillo, Enrique Deckler Colomer
技术建筑师：Jose Navarro
甲方：Council of the Alhambra and Generalife

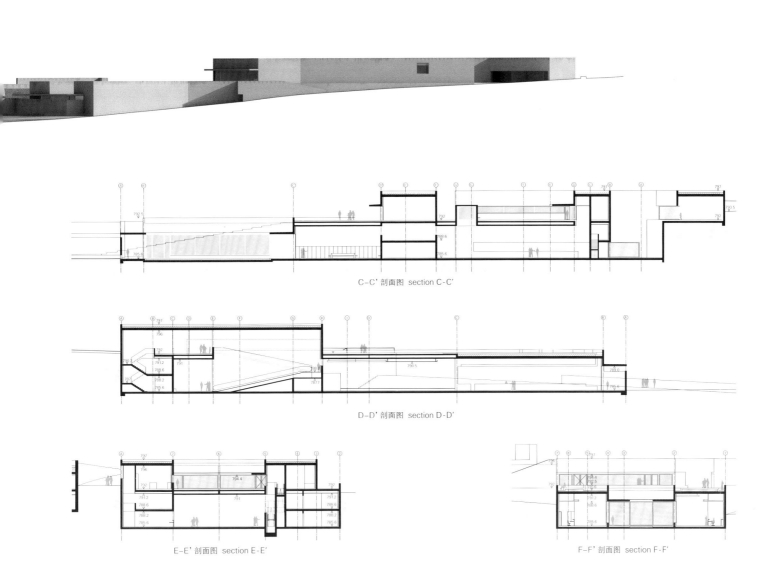

C-C' 剖面图 section C-C'

D-D' 剖面图 section D-D'

E-E' 剖面图 section E-E'

F-F' 剖面图 section F-F'

高塔VS天线 _Smiljan Radic + Gabriela Medrano + Ricardo Serpell

建筑师提议在山顶建一座综合型建筑，使山顶继续保留山顶的优势而不是被当做马鞍。它的形状介于高塔和天线之间：稳固、易辨识，另一方面又幻化成不稳定的意象。

这座综合建筑与其说像柱子，不如说更像虚幻的事物，因为它像柱子鬼影的骨架。从概念上来讲，它的形状并没有体现出面向未来的新颖性，因为其影像既不清晰也不易于即时理解，反而让人大惑不解。设计方案可谓集建筑师感兴趣的某些修复工作之大成，包括巴克敏斯特·富勒的张拉整体形式的不稳固性、肯尼思·斯内尔森的雕塑、维斯摩斯建构主义的一些尝试、亚历山大·罗辰寇空间多边螺旋结构的利用、纽万慧思采用的恒定塔模型，甚至包括当代的塞德里克·普若斯建造伦敦鸟舍时的环境，上述提及的匠心独运是这个建筑物体的真实来源。

建筑师相信多亏建筑设计关照了过去，复制于当下，才使这个富于全球化和都市化气息的分散型建筑结构不会昙花一现，瞬间过时。此外，它也无意与建于各个城市的标志物争奇斗艳，如果可能，仅希望是圣地亚哥试图在这个世界上树立一个属于自己的形象。塔结构是依据20世纪中期巴克敏斯特·富勒创立的张拉整体原则建成。在这种结构下，内部压缩的构件形成一张张力网，空间上却勾勒出系统，改变了人们对桁架结构的习惯性看法。

压缩的构件和伸张的构件间泾渭分明，可以清晰地解读主要压力并且展示动态的形式，这一切多亏了水平旋转角度的变化和构件形成的高度，尽管固定构件的长度有限。

Tower vs Antenna

The architects propose to build a hybrid object on the hilltop so that the hilltop maintains its character as a summit and not a saddle. Its appearance wavers be-

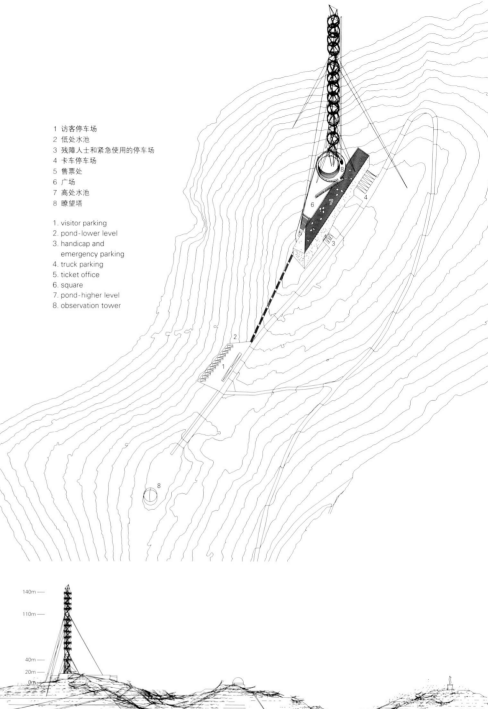

1 访客停车场
2 低处水池
3 残障人士和紧急使用的停车场
4 卡车停车场
5 售票处
6 广场
7 高处水池
8 瞭望塔

1. visitor parking
2. pond-lower level
3. handicap and emergency parking
4. truck parking
5. ticket office
6. square
7. pond-higher level
8. observation tower

A-A' 剖面图 section A-A'

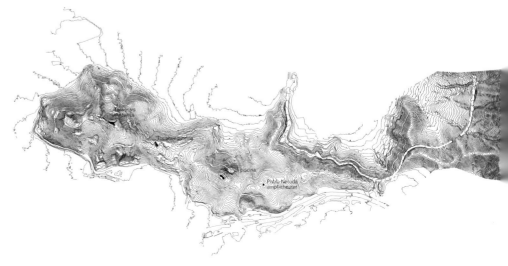

tween a tower and an antenna: between a stable, recognizable body, and one that dissolves into an unstable apparition. This hybrid is more like a ghost than a column. It is the skeleton of a ghost of a column. Conceptually, its shape does not propose any novelty for the future, as its figure is neither clear nor apprehensible at once in the present. Its reading is sharply confusing. Its form is derived from the recovery of certain works of architecture from the past that interests the architects. The formal instability of the Tensegrity structures

16

项目名称：Tower vs. Antenna
建筑师：Smiljan Radic, Gabriela Medrano, Ricardo Serpell
合作建筑师：Claudio Torres, Matias Valcarce
工程师：Pedro Bartolome
甲方：Government of Chile
设计时间：2014
施工时间：2014
竣工时间：2017

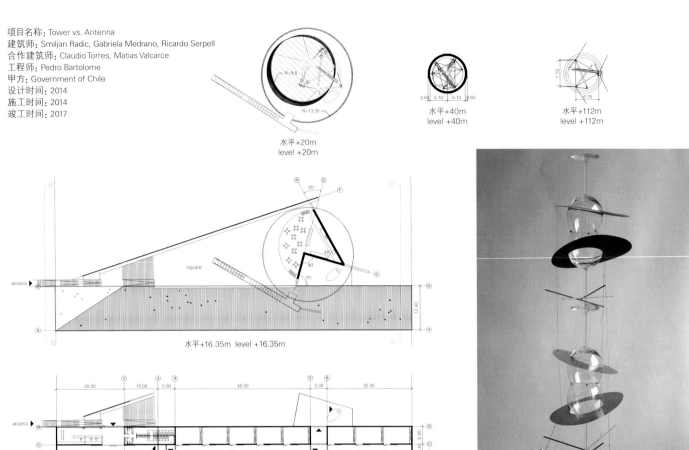

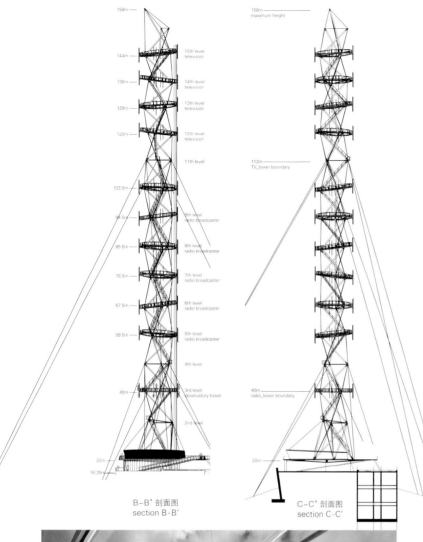

B-B' 剖面图
section B-B'

C-C' 剖面图
section C-C'

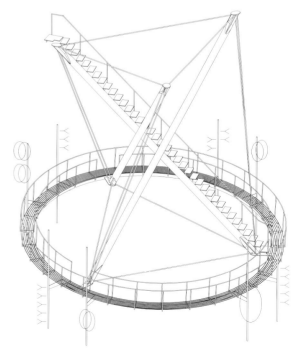

by Buckminster Fuller, the sculptures of Kenneth Snelson, some of the constructivists' exercises of Vuthemas, the polygonal spirals deployed in space by Aleksandr Rochenko, the handicraft of the Nieuwenhuys Constant tower maquettes, or, more recently, the environment of the Cedric Price's Aviary in London, all of them form the real memory of this object.

The architects believe that thanks to this retrospective view – to this apparent repetition – the globalized and diffuse form of this urban object will not be lost to the consumption of a quick spectacle, nor will it compete with the formal icons built in other cities, something desirable for a Santiago that tries to install its own image in the world.

The structure of the tower is based on the principle of tensional integrity or tensegrity developed by Buckminster Fuller by the mid-twentieth century. In this type of structure, members under compression are held inside a net of members in tension that spatially delineates the system, inverting the habitual perception of a trussed structure.

Compressed and tensioned members are explicitly distinguished, providing a clear reading of principal stresses and enabling dynamic formal possibilities thanks to the changes in the angle of horizontal rotation and in the height of the elements, despite using a limited set of fixed member lengths. Smiljan Radic + Gabriela Medrano + Ricardo Serpell

电子谷 _OMA

OMA建筑事务所为柏林的阿克塞尔·施普林格的新媒体中心而设计的作品在公共设计大赛的最终回合中获胜。建筑将向公众呈现阿克塞尔·施普林格媒体公司从"印刷出版"到"数字化"的跨跃过程,并在柏林中部现有的阿克塞尔·施普林格园区内创造一个新中心。

施普林格创立了一间由"印刷出版"跨越到"数字化"的公司——建筑在其中充当了一种象征和一个工具;像一座宫殿,吸引德国数字界的精英集聚在柏林。

印刷的优势之处在于其价格低廉、人工运作、曲高和寡,它是一个复杂的群策群力的过程,就这一点而言,迄今为止数字技术尚不能与之相提并论。建筑工作与报业的相似之处在于他们都需要对完全不同的信息来源进行复杂的汇编与选择。

按照建筑师的说法,办公室有着高效、精确和平稳的优势。但办公室也承受着一个严重的后果:劳动者与其计算机之间的关系,这种关系将其禁制在一个大气泡中独自含蓄地表演,难以参与到团队中。

传统的新闻编辑室被烟草、打字的记者所主宰,每个成员都能清晰地了解自己、同事和整个集体的工作及进展;同一个专题、同一个截稿期作为同一个解放日。在数字办公室,心无旁鹜地盯着屏幕抑制了对其他一切形式的关注,因此也潜在地破坏了要实现真正的创新所需的人工智慧。

因此,OMA建筑事务所提议建造这样一座大楼,大方地放送个人的工作成果用于共享分析。新办公大楼植入了一个中庭,面向既有的施普林格大厦开放——成为了施普林格园区的新中心。

设计方案的精髓是一系列阶梯状的楼层排列在一起形成一个数字谷。每层的一部分有屋顶覆盖作为传统的工作环境,到了平台部位便展露了出来。从侧面的半中腰处看过去,大楼形成了一个三维的天篷。

公众可以在三个层面上感受这座建筑——一楼大厅、会议桥和屋顶酒吧。会议桥是一个观景平台,游客可在此观看公司的日常事务是如何运作及如何演变的。

依靠平台连通形成的公共区域为大楼实体部分的正式办公区提供了替代选项,使得"工作区"这个词汇获得了空前的延伸;一座能够解开关于数字技术未来的一切问号的大厦。

Digital Valley

OMA's design for Axel Springer's new media center in Berlin has been chosen as the winner in the final round of a public design competition. The building will embody Axel Springer's corporate shift from print to digital media, and will create a new hub in the existing Axel Springer

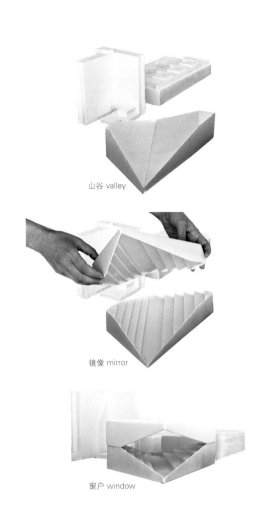

山谷 valley

镜像 mirror

窗户 window

Formal / Informal

Formal Office
75%
(25,728 m2)

Informal Office
25%
(8,576 m2)

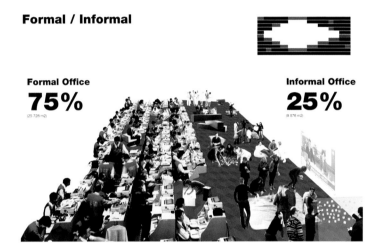

campus in central Berlin.

Springer has launched a corporate move from "print" to "digital media" – in which the building will act both as a symbol and a tool; a palace to lure the elite of Germany's digital area to its compound in Berlin. The genius of print is that it is a cheap, physical, hyper-accessible embodiment of a complex collective effort, for which so far the digital has been unable to find an equivalent. Architectural offices are similar to newspapers in that they produce complex assemblies and selections from radically different sources of information. According to the architect, they have experienced the advantages: speed, precision, smoothness. But they have also suffered one crucial consequence: the relationship between the worker and his computer, which isolates him in a bubble of introverted performance, inaccessible to collective overview.

In the classical newsroom, dominated by smoking, typing journalists, each inhabitant was aware of the labor and progress of his colleagues and of the collective aim: a single issue, with the deadline as a simultaneous release. In the digital office, staring intently at a screen dampens all other forms of attention and therefore undermines the collective intelligence necessary for true innovation.

Therefore, OMA proposes a building that lavishly broadcasts the work of individuals for shared analysis. The new office block is injected with a central atrium that opens up to the existing Springer buildings – a new center of the Springer campus.

The essence of the proposal is a series of terraced floors that together form a digital valley. Each floor contains a covered part as a traditional work environment, which is then uncovered on the terraces. Halfway through the building, the valley is mirrored to generate a three dimensional canopy.

The public can experience the building on three levels – ground floor lobby, meeting bridge, and roof-top bar. The meeting bridge is a viewing platform from which the visitors can witness the daily functioning of the company and how it evolves. The common space formed by the interconnected terraces offers an alternative to the formal office space in the solid part of the building, allowing for an unprecedented expansion of the vocabulary of workspaces: a building that can absorb all the question marks of the digital future.

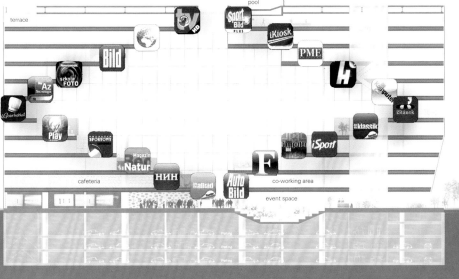

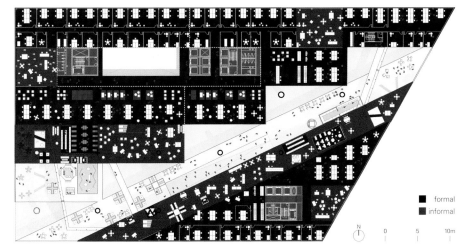

项目名称：Axel Springer Campus
地点：Axel-Springer-Straße 65, 10888 Berlin
建筑师：OMA
主要合作建筑师：Rem Koolhaas, Ellen van Loon
项目领导：Katrin Betschinger, Alain Fouraux, Betty Ng
项目团队：Anastasija Binevich, Anita Ernodi, Cindy Hwang, Claudio Saccucci, Denis Bondar, Edward Nicholson, Emile Estourgie, Frane Stancic, Gemawang Swaribathoro, Hans Larsson, Hendrik Hasenaar, Hyejun An, Jad Semaan, Janna Bystrykh, Jerome Picard, John-Paul Pacelli, Jonathan van Stel, Kostya Miroshnychenko, Lingxiao Zhang, Lam Le Nguyen, Marina Cogliani, Martin Murrenhoff, Martine Duyvis, Matthieu Boustany, Michael Hadjistyllis, Mike Yin, Minkoo Kang, Sara Bilge, Stefanos Roimpas, Tanner Merkeley, Tijmen Klone, Tom Shadbolt, Wai Yiu Man
结构、MEP、立面和可持续性：Arup London
微气候设计：RWDI / 音效设计：Kahle Acoustics
造价咨询：ARGE SMV Bauprojektsteuerung & Emproc GmbH
消防和安保：Peter Stanek
甲方：Axel Springer SE / 有效楼层面积：82,000m²

本地建筑的普适性_YongJu Lee

VERNACULAR VERSATILITY
CONTEMPORARY ADAPTATION OF KOREAN TRADITIONAL ARCHITECTURE

《evolo》杂志的摩天大楼设计大赛始于2006年,主要为了奖励突出的设计理念,通过创新性地利用技术、材料、功能、美学和空间结构来设计适宜居住的高层建筑。

2014年的大赛吸引了来自各大洲43个国家共计525件参赛作品,最终由建筑界领军人物成的评审团评出3位获奖者和20名优胜奖。第一名授予来自美国的李永居,他的参赛作品是"本地建筑的普适性"。设计方案重新诠释了韩国古建筑在当代多功能高层建筑中的使用。

"韩屋(HanOk)"是专用词,用来描述韩国古代房屋。韩屋带有裸露的木质结构系统和瓦片屋顶。屋檐的曲线结构可以调节射入室内的光线强度,而核心结构构件是一个叫做GaGu的木质连接结构。GaGu位于主屋顶体系的下方,即柱子和横梁及主梁交汇处,无需额外的组件,例如钉子,来固定结构,这个连接结构也是体现韩国古代建筑美学的一大特点。

历史上这个结构体系淋漓尽致地在平面上发展,但是仅限建造一层的住房。最近随着各种模型软件的问世,将这种传统工艺体系用于建造复杂的、能满足当代需要和项目要求的高层建筑的机会也随之增加。"本地建筑的普适性"这个项目开启了将旧建筑和传统设计高效且唯美地引入当今建筑的新篇章。

Vernacular Versatility
eVolo Skyscraper Competition was established in 2006 to recognize outstanding ideas for vertical living through the novel use of technology, materials, programs, aesthetics, and spatial organizations.
In 2014 competition, total of 525 projects from 43 countries in all continents were submitted and 3 winners and 20 honorable mentions were selected by the Jury, formed by leaders of the architecture.
The first place was awarded to YongJu Lee from the United States for his project "Vernacular Versatility". The proposal reinterprets traditional Korean architecture in

a contemporary mixed-use high-rise. HanOk is the name used to describe a traditional Korean house. A HanOk is defined by its exposed wooden structural system and tiled roof. The curved edge of the roof can be adjusted to control the amount of sunlight entering the house while the core structural element is a wooden connection named GaGu. The GaGu is located below the main roof system where the column meets the beam and girder and it is fastened without the need of any additional parts such as nails – this connection is one of the main aesthetic characteristics of traditional Korean architecture. Historically this structural system has been developed exclusively in plan, applied only to one-story residences. However, as various modeling softwares have been recently developed, there are more opportunities to apply this traditional system into complex high-rise structures that meet contemporary purposes and programs. Vernacular Versatility can open a new chapter of possibilities to bring this old construction and design tradition to the present day with efficiency and beauty.

项目名称：Vernacular Versatility
地点：Korea
建筑师：YongJu Lee
参与的竞赛：2014 eVolo Skyscraper Competition

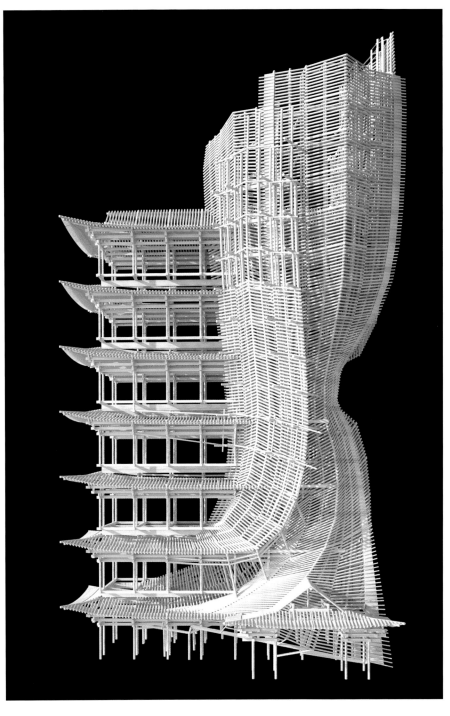

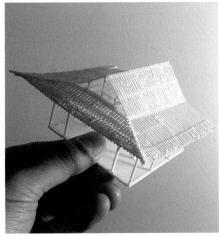

标准体系
standard system

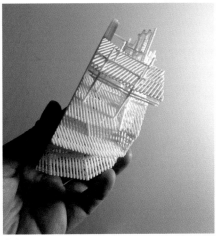

改造
transformation

新造型
new adaptation

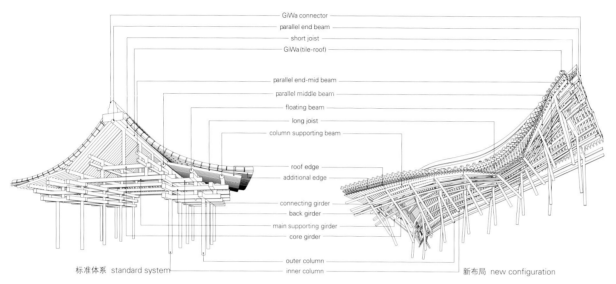

标准体系 standard system   新布局 new configuration

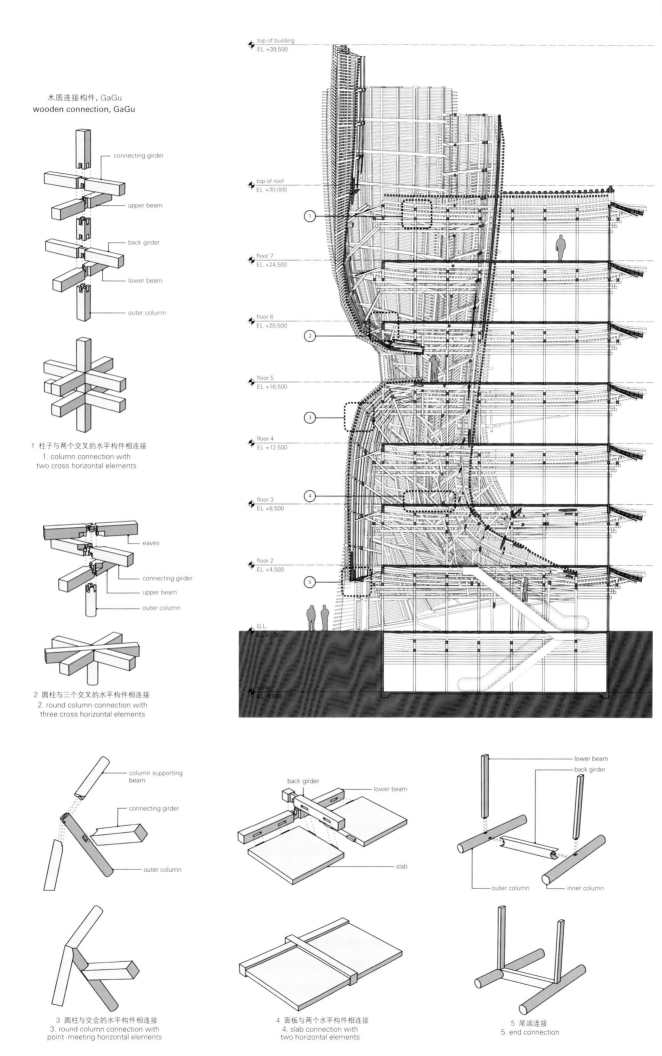

柳条剧院 _Tim Lai Architect + Brad Steinmetz Stage Design

这是一家临时的艺术设施和剧院,由金属管脚手架、园艺用起绒布、胶合板和可回收材料制成,尺寸为21m(长)×24m(宽)×6m(高)。

柳条剧院在2013年的世界舞台可持续剧院设计比赛中获奖,它建在威尔士卡迪夫的威尔士皇家音乐戏剧学院,用于为期十天的国际戏剧、歌剧和舞蹈节。作为重要的艺术设施和各种表演、讲座及会议场所,柳条剧院功能齐全,能够容纳150人。节庆结束时,所有材料保存起来并将得到重复利用。尽管剧院体积庞大,地上却无钻孔,场地可恢复原貌。

此项目由同在俄亥俄州哥伦比亚的建筑师Tim Lai和舞台设计师兼戏剧教授Brad Steinmetz合作完成。两位专家都对能够利用织物的多功能性和其友好亲近的特点来设计建筑物感兴趣。

柳条剧院的设计灵感来自于当地一种现成的叫做园艺用起绒布的材料,这种材料轻便,呈半透明状而且可以循环再用。多层的轻盈布条从天花板垂下来,整体建筑结构由模块化的脚手架构成。布条的形状和动作使人联想到摇摆的柳枝。当参观者朝着设有胶合板搭建成的阶梯长凳的剧院里面走时,可以与层层布条互动。剧院屋顶由不透明的池塘衬垫进行密封。晚上,柳条剧院会从里面透出光来。

Willow Theater

This is a temporary art installation and theater made of metal tube scaffoldings, horticultural fleece, plywood, and recyclable materials, sized 68'(d) × 80'(w) × 20'(h). Winning entry of World Stage Design for sustainable theater design competition 2013, Willow was constructed at the Royal

Welsh College of Music & Drama, Cardiff, Wales for the 10-day international festival of theater, opera, and dance. As a major art installation and venue for different performances, lectures, and conferences, the Willow is fully functional and can seat 150 people. When coming down at the end of the festival, all materials are saved and will be reused. Despite the large-size structure, no drilling to the ground was required. The site was restored to its original state.

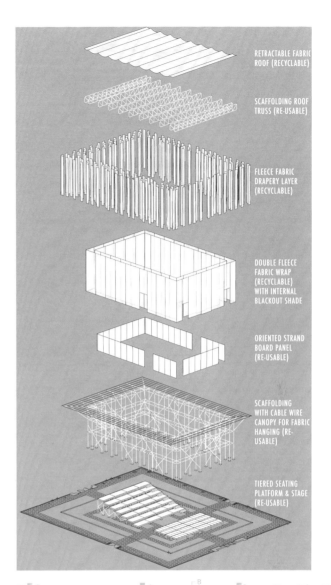

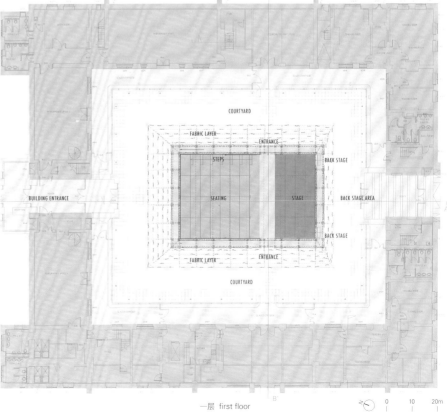

一层 first floor

Architect Tim Lai and stage designer and theater professor Brad Steinmetz, both based in Columbus, Ohio, collaborated on the project. Both professionals are interested in creating a structure that takes advantage of the versatility of fabric and its approachable qualities that are universally friendly.

The design of Willow was particularly inspired by a local and readily available material known as horticultural fleece, which is light-weight, translucent, and recyclable. Multiple layers of the airy fabrics hung from the ceiling and the structure was made of modular scaffoldings. The shape and motion of the fabric is reminiscent to the swaying branches of the willow tree. Viewers interact with layers of fabrics as they walk towards the interior, which is designed as a theater with step benches made of plywoods. The roof is sealed off with opaque pond liner. At night, Willow glows from within.

项目名称: Willow Theater
地点: Cardiff, Wales
建筑师: Tim Lai
舞台设计师: Brad Steinmetz
施工: Royal Welsh College of Music & Drama, Cardiff, Wales
甲方: World Stage Design
体积: 68'(d) x 80'(w) x 20'(h)m³
设计时间和竣工时间: 2013
摄影师: ©Matthew Carbone
(courtesy of the architect)

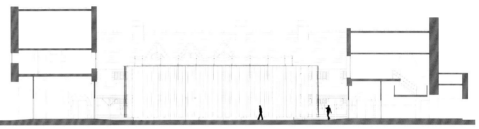
西立面 west elevation

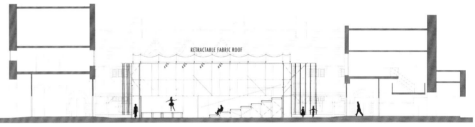
A-A' 剖面图 section A-A'

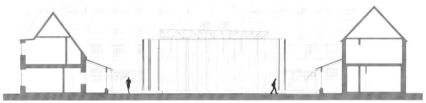
南立面 south elevation

B-B' 剖面图 section B-B'

Kineforum Misbar_Csutoras & Liando

Kineforum Misbar是一个临时的露天电影院,为2013年雅加达Kineforum双年展而建。Kineforum是鼓励艺术剧院和独立制片电影的一个非营利性组织。

该项目由社会文化议程推动。主要的目标是促进印度尼西亚电影的发展并创建一处各种背景的人们都能在大屏幕上观看电影的场所,包括那些没有能力在当地影院占优势的购物中心里的多厅影院里消费的人们。

参照目前在雅加达几乎绝迹的印尼的那些简陋、传统的露天影院,该影院起名为Kineforum Misbar,已于2013年12月免费播放电影。

项目位于雅加达市中心,印尼民族纪念碑莫纳斯(Monas)的基地处。38m×14m的影院周边覆着6m高的半透明"帷幕",使电影院置身在莫纳斯周边的大型开放区域中,并标出了内部空间的分界线。除了电影院本身,"帷幕"立面也包含一个宽敞的门厅,一个售票亭和一个快餐店。门厅空间被带有照明设备的天篷来界定,并且成为游客们在电影播放前后相遇和交往的氛围之地。立面的特色是沿着四周设有一条长凳,可供行人坐下休息,观看附近五人足球赛场的活动,晚上还能瞥见影院的演出。

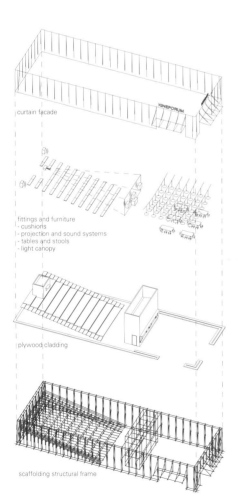

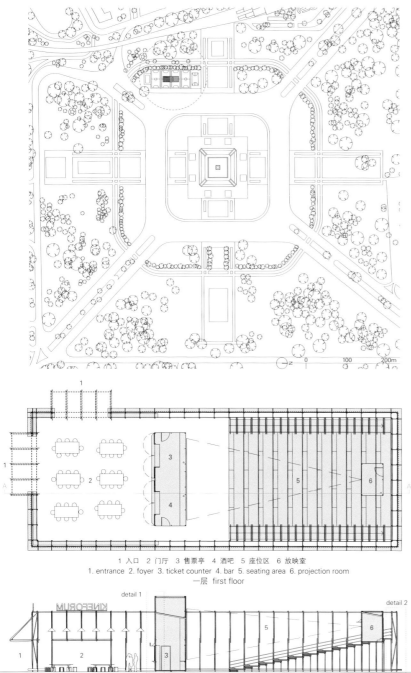

影院在10天内便建成了，使用了廉价的材料，且可重复利用。建筑结构是租用的脚手架管系统，适应性强，可快速组装。立面的帷幕材料名为agronet，是一种农业中常用的多孔织物。地面与墙壁都是由胶合板制成。墙壁由五颜六色的三角形构成的抽象图案所装饰，使表面更具活力，同时又能给参观者指引方向。灯罩采用白色的薄铝片灯罩制成，将室内照耀为白色，能更均匀地反射光线。

Kineforum Misbar

Kineforum Misbar is a temporary open-air cinema built as part of 2013 Jakarta Biennale for Kineforum, a non-profit organization promoting art-house and independent movies.

The project was driven by a social-cultural agenda. The main objectives were to promote Indonesian films and to create a venue where people of all backgrounds can watch a movie on the big screen, including those, who cannot afford the shopping mall's multiplexes that dominate the local cinema scene.

The cinema was named Kineforum Misbar in reference to the inexpensive, traditional Indonesian open-air cinemas which are now almost extinct in Jakarta and it screened movies for free in December, 2013.

1. 入口 2. 门厅 3. 售票亭 4. 酒吧 5. 座位区 6. 放映室
1. entrance 2. foyer 3. ticket counter 4. bar 5. seating area 6. projection room
一层 first floor

A-A' 剖面图 section A-A'

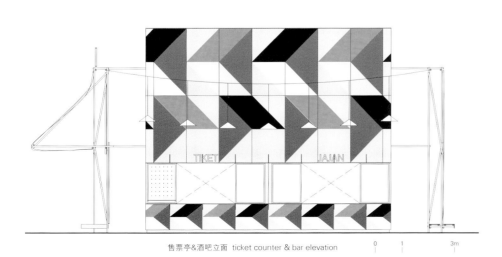

售票亭&酒吧立面 ticket counter & bar elevation

It was located in the center of Jakarta, at the base of Monas, Indonesia's national monument. The perimeter of the 38×14 meter structure was clad with a 6 meter tall translucent "curtain" to give the cinema presence in the large open area surrounding Monas and to define the boundaries of the spaces inside. Besides the cinema itself, the "curtain" facades also contained a generous foyer space and a pavilion housing a ticket counter and a snack bar. The foyer space was defined by a canopy of lamps overhead, which made it an atmospheric place for visitors to meet and mingle before or after a movie. The facade featured a bench along the perimeter, inviting passers-by to sit down and take a rest, watch the action on the neighboring futsal fields or, at night, take a peek at what's playing in the cinema. The building was constructed in 10 days, using inexpensive materials which were reused after the event. The structure was a rented scaffolding pipe system, which is highly adaptable and allows fast assembly. The material used for the facade curtain was agronet, a perforated fabric typically used in agriculture. Floors and walls were made out of plywood. The walls were decorated with an abstract pattern of colorful triangles which animated the surfaces and showed direction to the visitors. The lampshades were cut from thin aluminium sheets and sprayed white on the inside to more evenly reflect light.

项目名称: Kineforum Misbar
地点: Jakarta, Indonesia
建筑师: Csutoras & Liando
项目团队: Laszlo Csutoras, Melissa Liando
结构工程师: Sumarsono
用地面积: 530m² / 竣工时间: 2013
摄影师: courtesy of the architect

1. scaffolding base plate
2. scaffolding pipe
3. scaffolding swivel clamp
4. "agronet" curtain
5. tensioned steel cable
6. 40x40mm steel hollow section spacer
7. steel wire tie
8. 15mm moisture resistant plywood floor
9. white acrylic illuminated signage
10. 2x15mm moisture resistant plywood counter
11. 2x15mm moisture resistant plywood shutter/program display
12. corrugated steel roofing

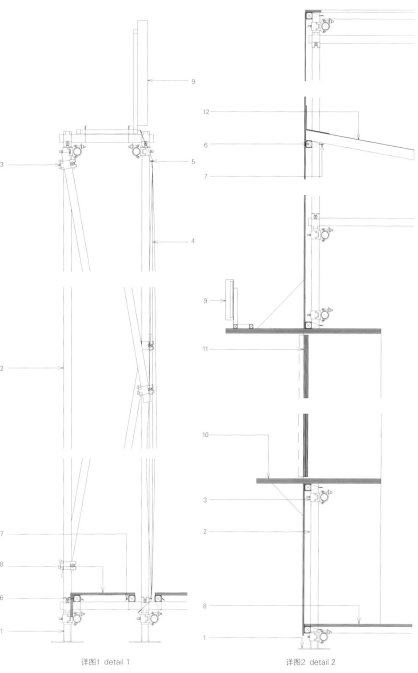

详图1 detail 1　　　详图2 detail 2

A Formal Tension

过去，形式的规则性、纯粹性、简单性曾是现代建筑的主要原则。而在最近几十年的建筑中，建筑师们不断扩大了现代建筑的形式词汇。另一方面，尤其是随着电子设计的出现，建筑师们已经开始尝试运用新的几何图形和施工技术，因此流动性、不规则性以及非标准的建筑形态已经成为我们现代都市风光的一部分。当这两种形式的建造方式在同一个建筑中并存时将会发生什么？自由与规则的建筑形态结合在一起是否会成为可能？还是取而代之，不可避免地形成冲突，产生富有张力的空间？本章节展示了两个项目——墨菲希思建筑师事务所设计的爱默生学院洛杉矶中心和Coop Himmelb(l)au设计的奥尔堡音乐之家。针对如上讨论，文章通过着重介绍自由和规则形态之间相互作用的区域，来为读者提供一种值得关注的见解。

In architecture of recent decades, architects have continued to elaborate the formal vocabulary of the Modern Movement, in which regularity, purity and simplicity of shape are among the main rules. On the other hand, and especially with the advent of digital design, architects have begun to experiment with new geometries and construction techniques, by which fluidity, irregularity and non-standard shapes have become part of our current cityscape. What happens when these two formal approaches co-exist in the same building? Is a marriage between free and regular shapes possible? Or does it instead inevitability create clash, generating areas of tension? This chapter presents two projects – the Emerson College Los Angeles Center by Morphosis Architects and the House of Music in Aalborg by Coop Himmelb(l)au – that may offer interesting insight into this discussion via a focus on areas of interaction between free and regular shapes.

爱默生学院洛杉矶中心_Emerson College Los Angeles Center/Morphosis Architects
奥尔堡音乐之家_House of Music in Aalborg/Coop Himmelb(l)au

形式的张力_A Formal Tension/Silvio Carta

我们可以在城市里很容易地找到从现代建筑师中习得的建筑语言的痕迹，其中纯粹的体量、立方体、平行六面体被看作是建筑的基本形态，通常在常规的设置中，曲线即使不是最主要的形态，也是同等重要的，例如，拱形和半圆（参见勒·柯布西耶的朗香教堂的礼拜堂，1954年，或者是TAC建筑事务所的Gropius的作品：巴尔的摩的Oheb Shalom寺，1957年，或者是路易斯·康的Dacca Assembly建筑，1962年）。

20世纪九十年代初，随着数字技术成为主流，建筑即使没有发生改革[1]，也发生了一次巨大的跳跃。这使得建筑从过去那些形式的约束条件中解放出来，例如次序、体量构成，以及后现代的象征主义。因此，现代建筑以非标准、非规范、不重复[2]为特征。UN工作室、NOX和格雷戈·林恩这些先锋设计师，他们在2003年巴黎蓬皮杜中心举行的非标准建筑展[3]中都展示了自己的作品，成为推动建筑"自由形态"的先锋。

然而，回顾过去二三十年的建筑，人们注意到无论是体量组成的建筑，还是自由形态的建筑，它们都没能够替代对方。相反，两者互相联系，这种联系以某种程度上无法解决的共存为特色。建筑师阿尔伯托·坎波·巴埃萨或者艾德瓦尔多·苏托·德·莫拉都以其纯粹、洁白、整齐的体量闻名于世，而其他一些建筑师如扎哈·哈迪德或弗兰克·盖里则以自由形态极具表现力的一面著称于世。我们的现代城市港湾都带有相似的公共外观，而自由形态和纯粹的体量都同样为大众所接受，两者产生的巨大冲突也成为当代城市建筑的新特色和公共形象。

在本章所分析的两个案例中，体量式的设计方式和自由形态并存，产生出一个形式混合体。在Coop Himmelb(l)au设计的音乐之家——丹麦奥尔堡的一个可容纳1300人的音乐厅和音乐学校中，我们可以清晰地看到波状的中央礼堂、U形学校和功能性建筑和谐共存。后者为音乐厅建立了巨大的水平框架，以城市的标准创造了清晰的室内和室外条件。学校利用规则的体量回归了城市，南向，可纵观整个峡湾。前院三座小型建筑延续了体量式的建筑，分别为私人厅、节奏厅和古典厅。规则的体量像是一个容器，定义了一处分明的空间，在物理上和视觉上建立了界

In our cities we can still easily find strong traces of the architectural language we inherited from the modern architects, in which pure volumes, cubes and parallelepipeds were considered basic shapes of composition. Curves were part of the equation as well, but often – if not predominantly – in a regular setting, such as in arches or semi-circumferences (think of Le Corbusier: The chapel of Notre Dame du Haut in Ronchamp, 1954; or Gropius by TAC: Temple Oheb Shalom in Baltimore, 1957; or Louis Kahn's Dacca Assembly Building, 1962).

A significant leap forward, if not a revolution[1], has occurred via the ascendance, beginning in the early 1990s, of digital design, which liberated architecture from formal constraints of the past, such as those of order, volumetric composition, and even Postmodern symbolism. The resulting architecture has been characterized as non-standard, non-normative, and non-repetitive.[2] Avant-garde designers such as UNStudio, NOX, and Greg Lynn, all of whom presented their work at the Non-Standard Architectures Exhibition[3] at the Pompidou Center in Paris in 2003, have been pioneers in promoting an architecture of "free shapes".

However, looking back on the architectural production of the last twenty, thirty years, one may observe that neither volumetric composition on the one hand nor freedom of shape on the other has actually displaced its opposite. Conversely, there is a relationship between the two that may be characterised as a sort of unresolved co-existence. Architects like Alberto Campo Baeza or Eduardo Souto de Moura are globally known for their pure, white and neat volumes, while others like Zaha Hadid or Frank Gehry have shown the expressive and performative aspects of free shapes all around the world. Our contemporary city harbor – with an apparently similar dose of public acceptance – both free shapes and pure volumes, validates a substantial clash of formal approaches that eventually becomes the new character and public image of such work.

In this chapter we analyse two cases in which a volumetric approach and free shapes cohabit, generating a formal mixture.

In Coop Himmelb(l)au's House of Music – a new 1,300-seat concert hall and music school in Aalborg, Denmark – we can clearly see the symbiosis generated by the central auditorium, with its weaving shape, and the U-shaped school and facility building. The latter creates a massive horizontal frame for the concert hall, establishing a clear inside-outside condition at the urban level. With its regular volume, the school building generates a back to the city, southward, opening up a frontward view of the fjord. The volumetric play continues in the front yard through three smaller buildings: the intimate hall, the rhythmic hall, and the classic hall. While the regular volumes act as a container, defining a clear space and creating limits both physically and visually, the concert hall appears continuous, flowing and open. The diagram

爱默生学院洛杉矶中心，加利福尼亚，美国，2014年
Emerson College Los Angeles Center, California, USA, 2014

奥尔堡音乐之家，丹麦，2014年
House of Music in Aalborg, Denmark, 2014

限。而音乐厅则是连续的、流动的、开放的。建筑师利用示意表来解释这方面的理念，使人一目了然：流向峡湾的、源自音乐厅中心的浅蓝色曲线以一种未定义的方式，从主广场和入口前方较小的功能性建筑之一开始，塑造了沿途所经的所有建筑的形态。这种流动的力量在礼堂的主要形态上有所体现，它遵循了声学逻辑，建筑师解释说："不定形的墙体灰泥结构和适应能力较强的顶棚旋吊体系设计均以声学专家的精确计算为基础，保证提供最佳的听觉体验。"另外，面向北方主入口的曲形天篷，连接入口（一个五层的中庭）和门厅的大型螺旋楼梯，以及建筑正面可以远眺峡湾的玻璃立面，都是流动空间的延续。

真正有趣的是两个常规体量与自由形态之间的互动。事实上，建筑师把交界区域作为支撑整座混合建筑物的重要手段。规则和自由形态交界所产生的空间成为展示"分享与协作"理念的接触点，由此产生的空间面向人们开放。建筑师解释说："通过这种设计，凡是能够共享的空间都成为公共空间、功能空间，而公共空间与功能空间之间的重叠区域的使用在整个设计中得以实现。"体量和音乐厅之间相互作用的区域既在

空间上成为公共空间，又在视觉上成为门厅。透过多层次的观察窗，学生和观众可以在门厅和培训室方向看到音乐厅，并能体验各种音乐活动，包括音乐会和排演。相互作用的空间能够调节建筑内部各个机构进行单独活动，同时允许使用者进行交流，将这种综合性建筑转变为一处生动、活跃的空间。

自由形态与体量的交互作用在墨菲希思建筑师事务所设计的位于好莱坞中心的爱默生学院洛杉矶中心（ELA）体现得更加深刻。建筑由顶部平台连接的两个十层结构以及两个交织的流动体量所占据的大型中央上空空间组成。塔楼主要供学生住宿，而中央的大型体量包含教室和行政办公室。体量曲线优美，其间的空间为使用者提供了公用空间。建筑师解释说："空间使不同楼层的露台充满了活力，同时促进了学生和行政人员的非正式交流。"在两个塔楼的内部，人们能找到学生公共教室的外部走廊。塔楼内部被一种带有纹理的的金属织物覆盖，延续了与内部庭院和露台的连接。项目中两个形态组合之间以冲突为基础的关系成为建筑物的主要特征之一。塔楼的造型像一个庞大的框架，流动的体量

the architects employ to explain this aspect of the concept is self-explanatory in this sense: Light blue curves generated from the centre of the concert hall branch off towards the fjord in an undefined way, shaping everything they encounter in their path, from one of the smaller performance buildings in the front to the main square and entrance. This fluid power is echoed in the main shape of the auditorium and follows acoustic logic: *"The design of the amorphous plaster structures on the walls and the height-adjustable ceiling suspensions, based on the exact calculations of the specialist in acoustics, ensures the optimal listening experience,"* explain the architects. Moreover, the fluid space continues in the curved canopy on the north-facing main entrance and in the spiral grand staircase which connects the entrance (a five-storey atrium) to the foyer and the main frontal glass facade overlooking the fjord.

What is truly interesting is the interaction between the two regular volumes and the free shapes. Such zones of interface are in fact utilised by the architects as a crucial device supporting the use of the overall compound. The spaces resulting from the encounter of rigid and free shapes become points of contact in which the concept of "sharing and synergy" unfolds. "Wherever possible sharing of spaces is enabled and overlapping of use of public and performance spaces is supported through the design," explain the architects. All the areas of interaction between the volumes and the concert hall become public spaces either physically, or the foyer, visually: "Through multiple observation windows, students and visitors can look into the concert hall from the foyer and the practice rooms and experience the musical events, including concerts and rehearsals." The interaction spaces regulate the individual activities of each institution within the compound and at the same time allow for interaction among the users, transforming the complex building into a lively and active place.

The interaction areas between free shapes and volumes receive even more emphasis in the Emerson College Los Angeles (ELA), located in the centre of Hollywood and designed by Morphosis Architects. The building consists of two 10-story structures connected by a platform at the top and a large central void occupied by two interweaving fluid volumes. While the towers are dedicated to student accommodations, the central volumes house classrooms and administration offices. The in-between spaces of the curvaceous volumes allow in their shapes common spaces for users, "activating terraces at different storeys and fostering informal interaction between students and staff," explain the architects. The inner side of the two towers – where one finds the exterior corridor of the student common rooms – is clad in a textured metal scrim, allowing continuous connection with the inner courtyard and the terraces. In this project the clash-based relationship between the two "families" of shapes becomes one of the main features of the building. The towers are shaped as an

1. Oxman Rivka "Theory and Design in the First Digital Age", The International Journal of Design Studies, 2006, 27(3): pp.229~265.
2. Ibidem.
3. Cf. http://www.designboom.com/contemporary/nonstandard.html

穿越其中。住宿部分越加的规则、严格，呈线形，教室和办公室则越反其道而行之。有趣的是，越是学院设置核心功能的区域，带有用于创作、创新和教育的空间，越是采用了动态形态。从这一点上看，我们可以注意到框架式形态的规则性被用来提供一种安全、朴素、坚固的形象，而交织的几何外形则给人不断更新进化的印象。在此，形态的运用展示了强烈的交流特征。混合型建筑展现了学校的一个概貌，反映了内部创新的动态起源的理念。这种描述可以做出多层次的解释。其中一个可能是对形式和教学理念做类比（"学生的教育背景和职业前景之间产生丰富的对话，并且以洛杉矶的地标建筑为背景来唤醒东海岸都市中心所积聚的能量"）。另外一种可能的解释则涉及剧院舞台的形象，建筑师解释说："在框架上运用大屏幕、媒体连接、声音和照明，上面的平台作为户外表演的灵活电枢，将起伏的织物转变为动态的视觉背景。整座建筑成为学生电影拍摄、放映和产业活动的舞台背景。"这两个项目阐明了两种处理自由形态与规则形态并存时产生的相互作用的空间的不同方法。在音乐之家项目中，自由形态和规则体量之间的相互作用仅仅通过主要建筑的变化的几何形态来感知，而在爱默生学院中，材料和形态的基础变化展现了这种相互作用。然而，建筑的共同特点是两种形态之间的不一致和互不联系。在这两个案例中，自由形态出现、突出并建立了与规则体量间的对立，而较为正式的结构则提供了一种坚固、冷静、静止的背景或整个建筑的框架。建筑师能够在同一个项目中运用如此不同的两种形式的建筑方式，得益于他们不但很好地理解了这种张力的存在，并且能将其作为一种强烈的表达特征运用到他们的设计中。因此，在同一座建筑中自由与规则形态的不相容所产生的冲突不但在解决功能问题（比如在同一座建筑中不同活动的相互作用）方面，而且在表达项目的形象方面都成为一种非常有效的手段。

enormous framework in which the fluid volumes pass through. The more regular, rigid and linear the residential part is, the more classrooms and offices seem to oppose to it. It is interesting to note that the more dynamic shapes are those where the core programme of the college takes place, with spaces dedicated to creativity, innovation and education. In this sense, one may observe that the regularity of the frame-like shapes is employed to provide an image of security, austerity and firmness, while the weaving geometries give the idea of something in continuous evolution and progression. Shapes have been employed here for their strong communicative characteristics. The compound confers a general picture of the School reflecting the idea of having a dynamic source of creativity inside. This description may allow for multiple interpretations. One may be the analogy between forms and teaching philosophy ("evoking the concentrated energy of East-coast metropolitan centers in an iconic Los Angeles setting, a rich dialogue emerges between students' educational background and their professional futures"). Another possible interpretation may refer to the image of the theater stage: "with screens, media connections, sound, and lighting incorporated into the framework, the upper platform serves as a flexible armature for outdoor performances, transforming the undulating scrim into a dynamic visual backdrop. The entire building becomes a stage set for student films, screenings, and industry events" explain the architects.

These two projects illustrate two distinct ways of dealing with the interaction spaces generated by the coexistence of free and regular shapes. While in the House of Music the interaction between the free shapes and the regular volumes is perceivable only through the changed geometry of the main buildings, in Emerson College the interaction is loudly announced by radical changes of materials and patterns. However, common to the two projects is a sense of disharmony and incommunicability between the two shape types. In both cases the free shapes emerge, stand out, and create contrast with the regular volumes, while the more formal structures provide a solid, calm and resting backdrop or frame for the overall composition. By using such different formal approaches in the same project, the architects appear not only to have understood very well the existence of such tension, but also to have taken advantage of it as a strong expressive characteristic of their design. Thus, the clash generated by the incompatibility of free and regular shapes in the same compound becomes here a powerful resource, not only in resolving functional questions (like the interaction between different activities in the same building), but also in expressing the new image of the project. *Silvio Carta*

爱默生学院洛杉矶中心
Morphosis Architects

爱默生学院总部设在马萨诸塞州波士顿,以传播和艺术课程而著名。爱默生学院洛杉矶中心(ELA)坐落在好莱坞中心地区,因此确定了学院在演艺业和美国第二大城市的中心地位。新中心坐落于日落大街,为爱默生学院现有的大学生实习项目建立了一个永久基地。该项目为那些学习传媒学院和艺术学院所开设的7门学科中任何一门的学生提供了ELA的实习机会。此外,ELA还提供研究生课程、学位课程和专业课程。新中心还将举办专题讨论会、讲座和其他与校友和洛杉矶社团相关的活动。

爱默生学院洛杉矶中心将学生住宿、教育设施和行政办公室集于一地,把一个大学校园的多样化凝缩在一个城市基地中。激发了东海岸城市中心在具有标志地位的洛杉矶所积聚的能量,使得学生的教育背景与他们的职业未来之间展开了内容丰富的对话。

作为爱默生学院洛杉矶中心的体验中最为基本的东西,学生的居住环境是整座建筑的主要部分。它能为217个学生提供住宿,其内部区域构成了一个动态核心,致力于创造、学习和社会交往。两栋细长的住宅塔楼由一个多功能平台连接起来,10层高的方形框架围出了一个中心开阔的体量,创建了一处灵活的户外"房间"。用作教室和行政办公室的雕刻形式的建筑体蜿蜒穿过这个体量,形成了多层露台,以及用于开展非正式社会活动和创造性思想交流的、活跃的缝隙空间。人们可以远眺多功能露台,并且看见通往学生套间和公共休息室的外部走廊由跨越塔楼内部表面的整个高度的波状、带有纹理的金属织物来进行遮阴。

留意当地环境,中心以好莱坞电影工作室的内部结构为范例,即建筑的常规外立面设有灵活、奇妙的内部空间。屏幕、媒体链接、音响和照明都融入到框架中,上面的平台充当户外表演的灵活支架,将波状织物变成了一个动态的视觉背景。整座建筑成了一个舞台场景,用于学生电影、放映和行业活动,不但有着好莱坞的标志,而且洛杉矶市和远处的太平洋更增添了一幅景致。

为了获得LEED金奖评级,新中心支持爱默生中心的可持续设计和社会责任。建筑的正面为东西朝向,居住塔楼以动态的室外表皮为特色。为了适应当地的天气条件,自动化的遮阳系统可开启和关闭高性能玻璃幕墙外的水平翅片,以在最大程度地获得日光和风景的前提下将热量的获得减到最小。更深层次的绿色倡议包括使用可循环和快速再生的建筑材料;安装有效装置,以将水的使用减少40%;通过无源价系统进行加热和制冷时节省能源;建筑的管理以及用于监控和优化所有系统有效性的基础设施试运行。

Emerson College Los Angeles Center

Based in Boston, Massachusetts, Emerson is renowned for its communication and arts curriculum. Located in the heart of Hollywood, Emerson College Los Angeles(ELA) defines the College's identity in the center of the entertainment industry and the second largest city in the United States. The new facility establishes a permanent home on Sunset Boulevard for Emerson College's existing undergraduate internship program that will extend the ELA experience to students studying in any of the seven disciplines that are offered through the School of Communication and the School of the Arts. Additionally, ELA will offer post-graduate, certificate, and professional study programs. The new facility will also host workshops, lectures, and other events to engage with alumni and the LA community.

Bringing student housing, instructional facilities and administrative offices to one location, ELA condenses the diversity of a college campus into an urban site. Evoking the concentrated energy of East-coast metropolitan centers in an iconic Los Angeles setting, a rich dialogue emerges between students' educational background and their professional futures.

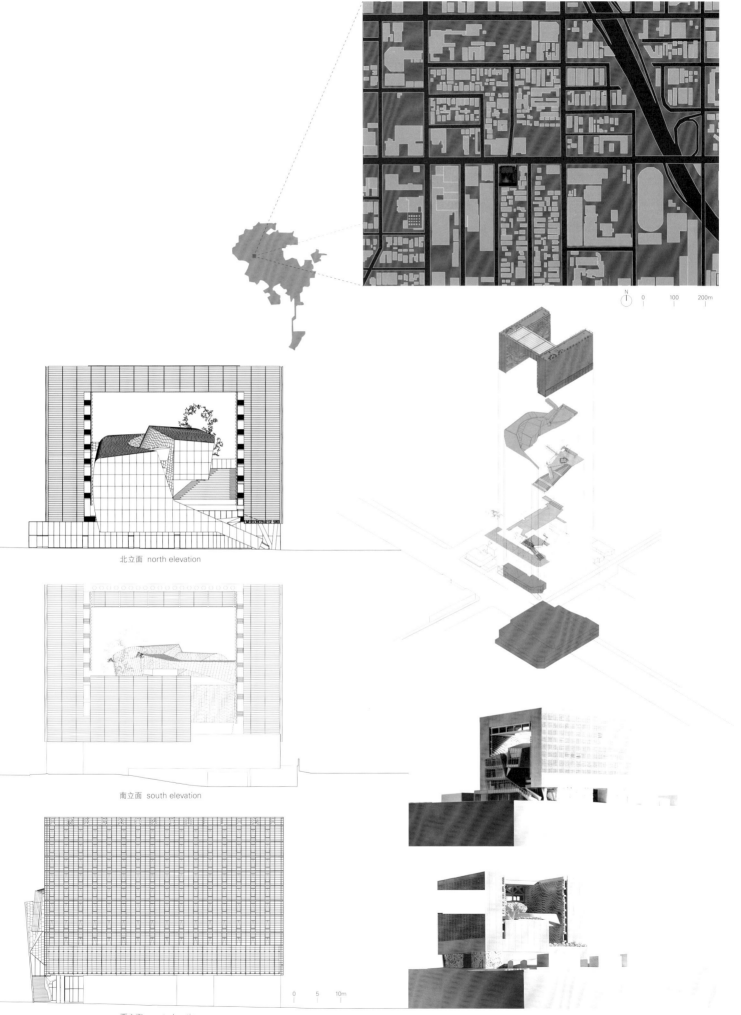

北立面 north elevation

南立面 south elevation

西立面 west elevation

Fundamental to the ELA's experience, student's living circumstances give structure to the overall building. Housing up to 217 students, the domestic zones frame a dynamic core dedicated to creativity, learning, and social interaction. Composed of two slender residential towers bridged by a multi-use platform, the 10-story square frame encloses a central open volume to create a flexible outdoor "room". A sculpted form housing classrooms and administrative offices weaves through the void, defining multi-level terraces and active interstitial spaces that foster informal social activity and creative cross-pollination. Looking out onto the multi-level terrace, exterior corridors to student suites and common rooms are shaded by an undulating, textured metal scrim spanning the full height of the towers' interior face.

Looking to the local context, the center finds a provocative precedent in the interiority of Hollywood film studios, where outwardly regular facades house flexible, fantastical spaces within. With screens, media connections, sound, and lighting incorporated into the framework, the upper platform serves as a flexible armature for outdoor performances, transforming the undulating scrim into a dynamic visual backdrop. The entire building becomes a stage set for student films, screenings, and industry events, with the Hollywood sign, the city of Los Angeles, and the Pacific Ocean in the distance providing added scenery.

Anticipated to achieve a LEED Gold rating, the new center champions Emerson's commitment to both sustainable design and community responsibility. Defining the building's facades to the East and West, the residential towers feature an active exterior skin. Responding to local weather conditions, the automated sunshade system opens and closes horizontal fins outside the high-performance glass curtain-wall to minimize heat gain while maximizing daylight and views. Further green initiatives include the use of recycled and rapidly renewable building materials, installation of efficient fixtures to reduce water use by 40%, energy savings in heating and cooling through a passive valence system, and a building management and commissioning infrastructure to monitor and optimize efficiency of all systems.

项目名称：Emerson College Los Angeles Center
地点：Los Angeles, California, USA
建筑师：Morphosis Architects
设计总监：Thom Mayne
项目负责人和经理：Kim Groves
首席项目设计师：Chandler Ahrens
项目建筑师：Aaron Ragan
项目设计师：Shanna Yates
项目团队：Natalia Traverso Caruana, Brock Hinze, Yasushi Ishida, Jai Kumaran
项目助理：Katsuya Arai, Marco Becucci, Chris Bennett, Cory Brugger, Amaranta Campos, Joe Filippelli, Alex Fritz, Penny Herscovitch, Hunter Knight, Zachery Main, Jon McAllister, Nicole Meyer, Cameron Northrup, Brandon Sampson, Michael Smith, Scott Smith, Ben Toam, Elizabeth Wendell
结构：John A. Martin Associates, Inc.
MEP：Buro Happold
景观建筑师：Katherine Spitz Associates
立面：JA Weir Associates
土工技术：Geotechnologies
照明：Horton Lees Brogden Lighting Design, Inc.
音效：Newson Brown Associates LLC
视听设计：Waveguide Consulting Inc.
总承包商：Hathaway Dinwiddie Construction Company
用地面积：3,000m² / 面积：11,148m²
设计时间：2008—2010 / 施工时间：2012—2014 / 竣工时间：2014
摄影师：©Iwan Baan(courtesy of the architect)(except as noted)

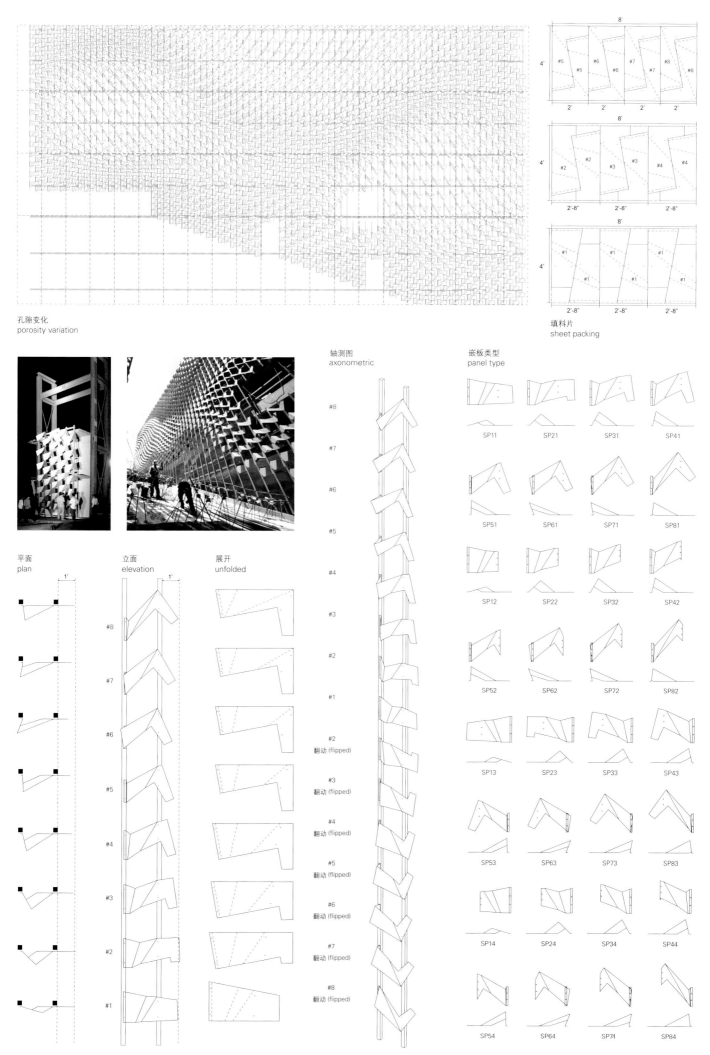

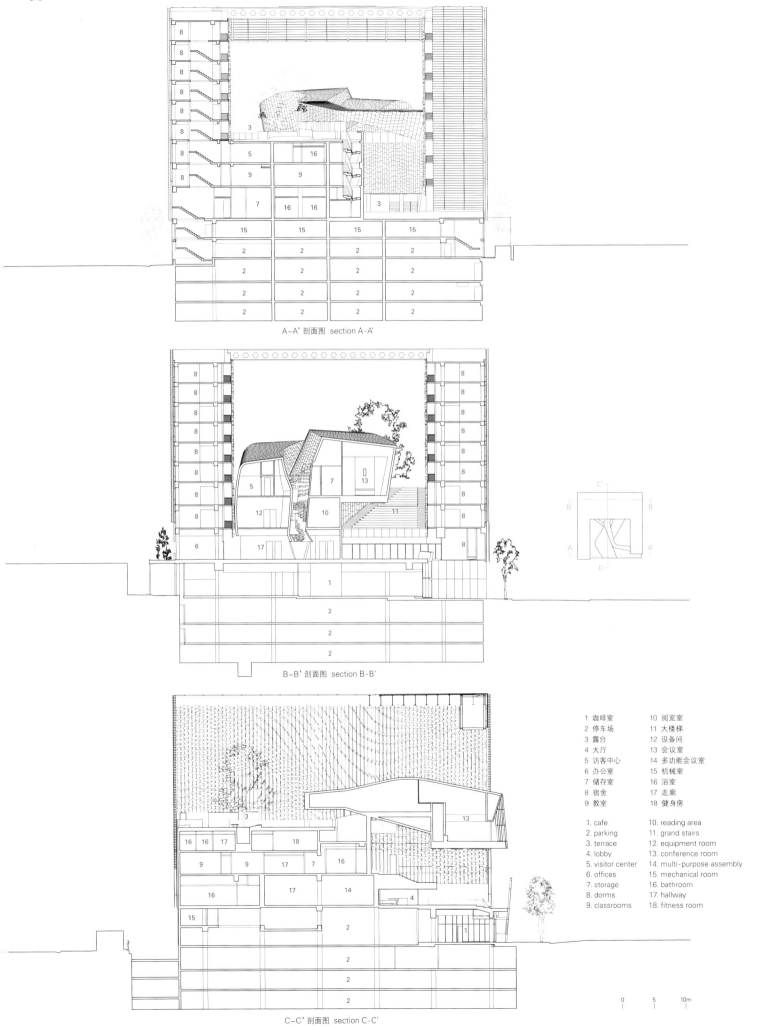

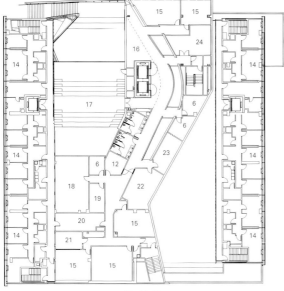
三层 third floor

1 咖啡室	10 Vin Di Bono远程	18 计算机实验室
2 停车场	教学室	19 听力实验室
3 露台	11 访客中心	20 编辑实验室
4 工作室	12 存储室	21 研讨室
5 大厅	13 音频后期制作室	22 控制室
6 办公室	14 宿舍	23 计算机室
7 更衣室	15 教室	24 设备间
8 明亮的放映室	16 阅览室	25 会议室
9 多功能会议室	17 大楼梯	26 公共休息室&厨房

1. cafe	10. Vin Di Bono distance	18. computer lab
2. parking	learning room	19. audio lab
3. terrace	11. visitor center	20. editing lab
4. studio	12. storage	21. seminar room
5. lobby	13. audio post	22. control room
6. offices	14. dorms	23. server room
7. dressing rooms	15. classrooms	24. equipment room
8. bright screening room	16. reading area	25. conference room
9. multi-purpose assembly	17. grand stair	26. common rooms & kitchen

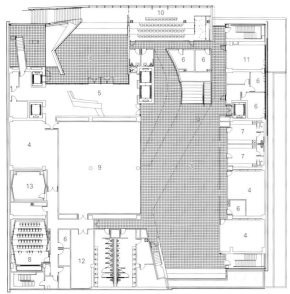
二层 second floor

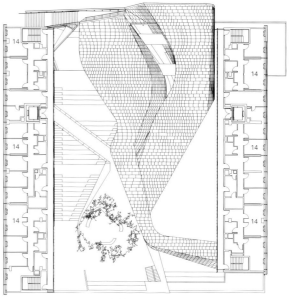
七层 seventh floor

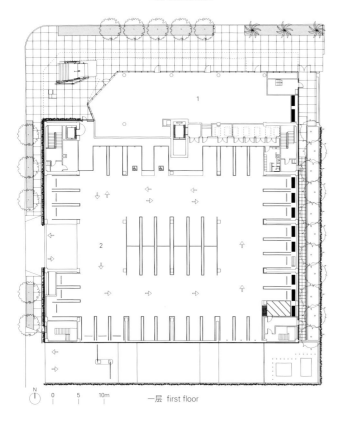
一层 first floor

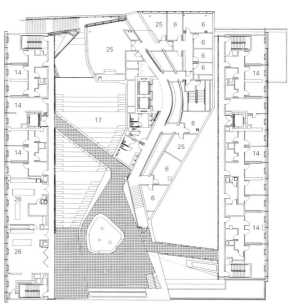
五层 fifth floor

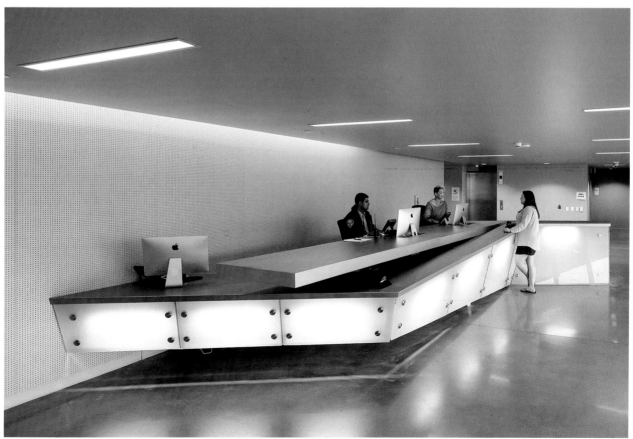

奥尔堡音乐之家
Coop Himmelb(l)au

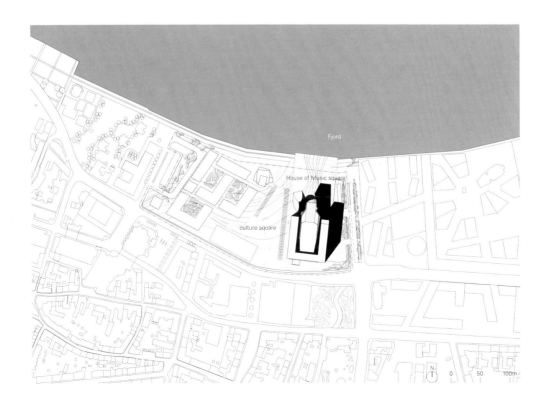

Coop Himmelb(l)au建筑工作室的设计将文化中心的音乐学校和音乐厅合并为一个整体：开放式的结构设计促进了观众与艺术家、学生与老师之间的交流。Coop Himmelb(l)au工作室CEO、首席设计师Wolf D. Prix解释说："建筑背后的理念可以通过外部造型表现出来。音乐学校环绕着音乐厅。"

U形排练&培训室围绕作为建筑核心的音乐厅（可容纳1300名观众）建立。宽敞的门厅连接了这些空间，并且通过多层的窗口面向附近的文化空间和一个峡湾开放。门厅下方是私人厅、节奏厅和古典厅，这三个不同规格的厅馆对空间进行了补充。通过多层的观测窗口，学生和访客从门厅和练习室的方向能够看到音乐厅，体验音乐会和排练这样的音乐活动。

音乐厅

礼堂内部流动的曲线造型与外部棱角分明的立方体体量形成鲜明对比。管弦厅和曲形阳台内的座位设计能为听众带来最好的声音效果和视觉效果。高品质的混合声响理念是工作室与奥雅纳的Tateo Nakajima合作开发的。墙体上不定形的灰泥结构和适应能力较强的顶棚旋吊体系设计均以声学专家的精确计算为基础，保证提供最佳的听觉体验。音乐厅的降噪等级为NR10 (GK10)，使这里成为欧洲最安静的场馆之一。音乐厅凭借其建筑特色和声学品质早已被预定：2014年4月，人们在这里将能欣赏到英国皇家爱乐乐团与小提琴演奏家Arabella Steinbacher，以及丹麦国家交响乐团与女高音歌唱家Mojca Erdmann的演奏。

门厅

门厅是为学生、艺术家、老师和观众提供的集会场所。人们透过大窗户可俯视整个峡湾。因此这个带有楼梯和瞭望阳台的五层高的门厅是一个用于各种活动的动态空间，充满活力。

Wolf D. Prix说，"音乐之家"是音乐与建筑相结合的一个标志。"音乐是能够直接拨动人们心弦的艺术。"与乐器类似，音乐之家是对创新性的一个很好的回应。

能源理念

门厅利用较大的垂直空间内自然上升的热气流来代替风扇，进行通风。混凝土楼板下铺设的水管用于夏季制冷和冬季制暖。围绕音乐厅的混凝土墙是额外的热能存储系统。峡湾则提供了免费的制冷动力。

音乐厅的管道系统和通风系统采用高效的旋转式热交换器。音乐厅座位的下方附有可使气流低速流动的高效通风系统。

空气通过照明系统上方的天花板网格排出，因此室内产生的任何热量都不会使气温升高。

建筑管理规划控制着建筑内部的所有设施，确保任一系统在不必要的情况下是关闭的。能源消耗通过这种方式达到最小化。

House of Music in Aalborg

This cultural centre was designed by the Viennese architectural studio Coop Himmelb(l)au as a combined school and concert hall: its open structure promotes the exchange between the audience and artists, and the students and teachers. *"The idea behind the building can already be read from the outer shape. The school embraces the concert hall"*, explained Wolf D. Prix, design principal and CEO of Coop Himmelb(l)au.

U-shaped rehearsal and training rooms are arranged around the core of the ensemble, a concert hall for about 1,300 visitors. A generous foyer connects these spaces and opens out with a multi-storey window area onto an adjacent cultural space and a fjord. Under the foyer, three rooms of various sizes complement the space: the intimate hall, the rhythmic hall, and the classic hall. Through multiple observation windows, students and visitors can look into the concert hall from the foyer and the practice rooms and experience the musical events, including concerts and rehearsals.

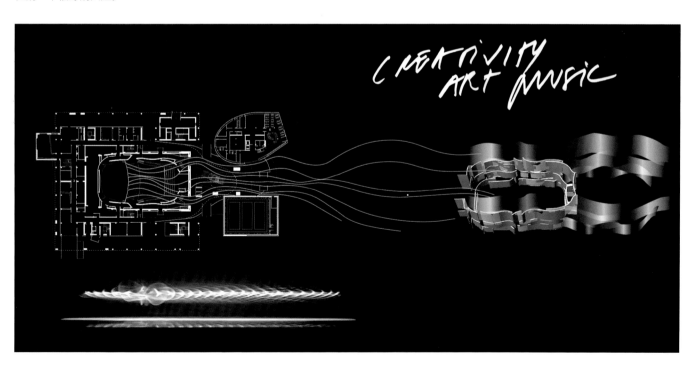

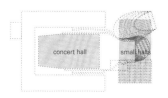
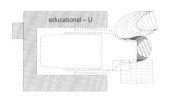
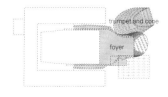
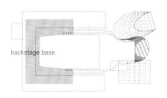

The concert hall

The flowing shapes and curves of the auditorium inside stand in contrast to the strict, cubic outer shape. The seats in the orchestra and curved balconies are arranged in such a way that offers the best possible acoustics and views of the stage. The highly complex acoustic concept was developed in collaboration with Tateo Nakajima at Arup. The design of the amorphous plaster structures on the walls and the height-adjustable ceiling suspensions, based on the exact calculations of the specialist in acoustics, ensures for the optimal listening experience. The concert hall will be one of the quietest spaces for symphonic music in Europe, with a noise-level reduction of NR10 (GK10). Thanks to its architectural and acoustic quality, the concert hall is already well-booked: there will be concerts featuring the Royal Philharmonic Orchestra with violin soloist Arabella Steinbacher and the Danish National Radio Orchestra with soprano Mojca Erdmann in April.

The foyer

The foyer serves as a meeting place for students, artists, teachers, and visitors. Five stories high with stairs, observation balconies, and large windows with views of the fjord, it is a lively, dynamic space that can be used for a wide variety of activities.

For Wolf D. Prix, the "House of Music" is a symbol of the unity between music and architecture: "Music is the art of striking a chord in people directly. Like the body of musical instruments this architecture serves as a resonance body for the creativity in the House of Music."

The energy concept

Instead of fans, the foyer uses the natural thermal buoyancy in the large vertical space for ventilation. Water-filled hypocaust pipes in the concrete floor slab are used for cooling in summer and heating in winter. The concrete walls around the concert hall act as an additional storage capacity for thermal energy. The fjord is also used for cost-free cooling.

The piping and air vents are equipped with highly efficient rotating heat exchangers. Very efficient ventilation systems with low air velocities are attached under the seats in the concert hall.

Air is extracted through a ceiling grid above the lighting system so that any heat produced does not cause a rise in the temperature in the room.

The building is equipped with a building management program that controls the equipment in the building and ensures that no system is active when there is no need for it. In this way, energy consumption is minimized.

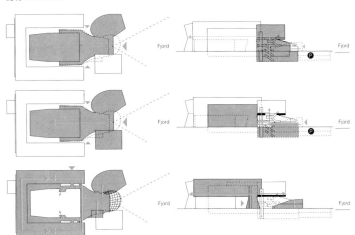

流线 circulation

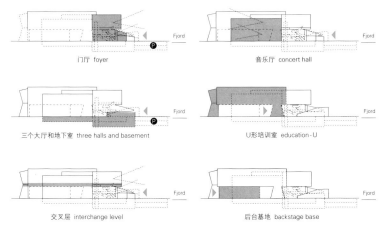

布局 organization

门厅 foyer　　音乐厅 concert hall
三个大厅和地下室 three halls and basement　　U形培训室 education-U
交叉层 interchange level　　后台基地 backstage base

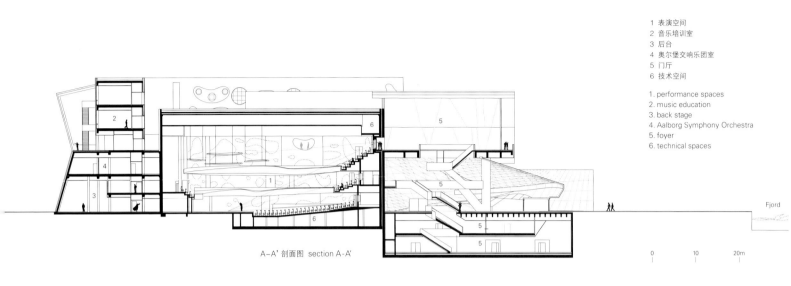

A-A' 剖面图 section A-A'

1 表演空间
2 音乐培训室
3 后台
4 奥尔堡交响乐团室
5 门厅
6 技术空间

1. performance spaces
2. music education
3. back stage
4. Aalborg Symphony Orchestra
5. foyer
6. technical spaces

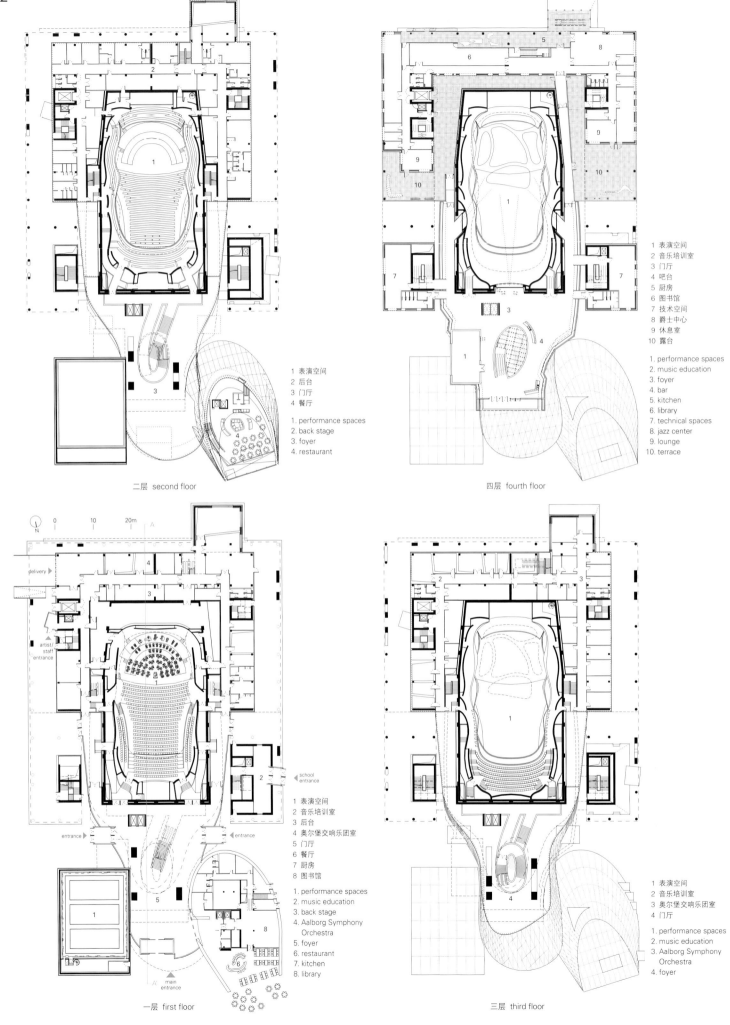

项目名称：House of Music II / 地点：Aalborg, Denmark
建筑师：Coop Himmelb(l)au / 首席设计师：Wolf D. Prix
合作建筑师：Michael Volk / 设计建筑师：Luzie Giencke
项目建筑师：Marcelo Bernardi, Pete Rose
室内设计建筑师：Eva Wolf
项目团队：Markus Baumann, Wendy Fok, Robin Heather, Ivana Jug, Ariane Marx, Anja Sorger, Bo Stjerne Thomsen, Anna Wasserthal, Philip Wilck, Blaine Lepp, Talya Kozminsky, Hannes Walzl, Morten Grau Jensen, Stephanie Neufeld, Tyler Bornstein, Laura Githa, Julia Gärtner, Jenny Draxlbauer
CAD协调员：Ronny Böser, Benjamin Schmidt
3D可视化系统：Armin Hess/Isochrom
本地建筑师：Friis & Moltke
音效、视听&剧院设计、规划顾问：Arup
景观建筑师：Jeppe Aagaard Andersen
结构工程师：Rambøll, B+G Ingenieure
机械&电气、防火工程师：Nirás
成本顾问：Davis Langdon LLP
照明设计顾问：Har Hollands
室内设计顾问：Eichinger Offices
甲方：North Jutland House of Music Foundation
总建筑面积：17,637m² / 有效楼层面积：20,257m²
设计时间：2003 / 施工时间：2010 / 竣工时间：2014
摄影师：©Martin Schubert(courtesy of the architect) (except as noted)\

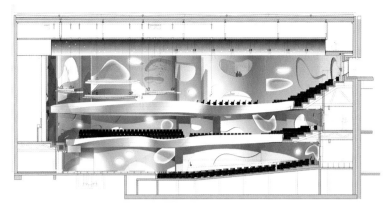

起居与工作相融合
Live/Work

路易斯·芒福德在《城市发展史》一书中指出,中世纪的起居和工作的融合空间在现代城市中仅停留在画家和建筑师的工作室中。以仅有200年历史的办公楼为典型,它已经成为现代城市中最杰出的建筑类型之一。随着自动化持续地减少体力劳动,以及城市化进程的加速,未来我们在何种工作空间中工作将非常重要。

至少长达50年的技术进步已经促进了办公室将被淘汰这一预言的实现,但是人们仍然顽固地需要聚到一起,近距离地工作,但是21世纪通讯技术的发展最终会使办公室解体成为可能吗?在这里我们将看到大量的小型办公室项目,由此提出关于未来工作空间将如何布局的问题。

Lewis Mumford, in *The City in History*, noted that the medieval synthesis of living and working space only remained in the modern city through the studios of painters and architects. The office block, barely 200 years old as a typology, has come to be one of the most prominent types of building in the contemporary city. With automation continually diminishing manual labour, and with urbanization still accelerating, what kind of office space we work in future will be incredibly important.

For at least 50 years technological advancement has encouraged the prediction that the office will become obsolete, but there is still a very stubborn need for people to come together and work in close proximity. But does the advance of communication technology in the 21st century finally make the dissolution of the office possible? Here we look at a number of small office projects that raise questions about the way working space will be organised in future.

SISII花园式展示厅_SISII/Yuko Nagayama & Associates
Koen Van Den Broek工作室_Koen Van Den Broek Studio/Haerynck Vanmeirhaeghe Architecten
Draft工作室与住宅_Draft Studio and House/Estudio Elgue
绿色工作室_The Green Studio/Fraher Architects
地下车库建筑工作室_Underground Garage Architecture Office/Carlo Bagliani
稻田中的办公室_Office in Rice Field/Akitoshi Ukai/AUAU
Roduit工作室_Roduit Studio/Savioz Fabrizzi Architectes

起居与工作相融合_Live/Work Hybrids/Douglas Murphy

Hybrids

自20世纪中期，城市中大多数最著名的建筑就是办公室。

当时宗教和皇室建筑一度占据了地平线，作为（政治和精神方面的）权力的聚焦点而存在，自从发明了钢框架和升降电梯以后，摩天办公大楼成为主导。然而这些建筑体现的权力是世俗的，也是与经济相联系的。今天的高楼很高因为这些高楼建造在有价值的土地上，从而能得到多种经济回报。这些摩天大楼不管是有意的还是无意的，都标记了今天世界上势力存在的位置。

办公室是近代的人类发明之一，为非体力劳动力提供不同的空间，一直到19世纪初期，宗教或政府之外的环境是全然未知的。在工业革命时期以前，在中世纪的欧洲，几乎没有办公建筑，工厂是一处房屋，在那里工匠、他们的家人和员工在一起生活和工作，家庭和工作生活之间还没有被严格地区分。

欧洲工业化于18世纪开始，工匠作坊突然扩大规模成为工厂，同时相应的商业发展和贸易法律使特定空间成为必要，保险和银行的人们在那里能够工作。但是直到19世纪，工业化世界中资本主义呈优势发展，专门为行政和牧师的工作而设计的建筑开始占据了城市中的典型区域。

随着美国升至世界主导地位，由钢铁和玻璃建造的美国办公大楼和摩天大厦展现了资本主义现代化的高度。机械化服务的优化配置允许设置更大的内部区域，使现代开放式办公室得以发展，细分成超高效的网格、同等的照明、同等的通风和不断增加的抽象的空间体验。

在20世纪60年代快速的技术变革的背景下，将办公室想象为无限空间内不断重构的模型成为可能。然后，在20世纪70年代，像Norman Foster这样的建筑师设计的办公室是在一处简单的服务空间内，简单地覆盖了场地，与边缘右对齐，仅利用玻璃窗帘墙进行封闭。此举创造了当代办公室的参数，这些参数是高度服务型且计算化的，本质上在全球是相同的。

但是人们的工作方式一直在变化。开放式办公室周期性地流行和过时，如"办公室景观"的创新、"办公桌轮用制"的花招。但是由于全球化网络基础设施的发展，更大的改变似乎正在进行中。在家工作是一个长久讨论的话题，但是人们离开办公室工作的前景与当代对"合作"的热情相抵消，而在"合作"中极为贴近的其他思想被认为是创造和创新链中的必要连接。

Since the middle of the 20th century, frequently the most prominent buildings in a city are its offices.
Where once religious and royal buildings dominated the skyline, marking as focus points the locations where power – both political and spiritual – resided, the invention of the steel frame and the elevator, the skyscraping office block has come to dominate. The power that these buildings mark out, however, is secular, economic. The tall building today is tall because it is financially prudent to multiply the returns from building on a valuable land. That these skyscrapers also mark out the location where power resides in today's world may or may not be deliberate.
The office is one of the more recent human inventions. The provision of distinct space for non-manual labour, outside of a religious or governmental context was almost unknown as late as the beginning of the l9th century. In the pre-industrial era there was little architecture devoted to "work" – in medieval Europe the workshop was a house where the artisan, their family and their staff would all live and work together, and the strict division between family and work life was yet to become so delineated.
As industrialisation took off in Europe in the 18th century, the artisan workshop swelled to become the factory, while the corresponding development of commerce and the laws of trade necessitated the creation of specific spaces in which insurance and banking could take place. But it wasn't until the l9th century and the growing dominance of capitalism in the industrialised world that buildings specifically designed for administrative and clerical work began to take their place in the city as a typology distinct in themselves.
As America rose to its dominant world position, so the American office block, skyscraping, built of steel and glass, came to represent the height of capitalist modernity. The increasing deployment of mechanical services allowed for greater internal areas, leading to the development of the modern open-plan office, subdivided into super-efficient grids, equally lit, equally ventilated, an increasingly abstract experience of space.
In the context of the rapid technological change of the 1960s, it became possible to image the office as a model for infinite space – endlessly reconfigurable. Then, in the 1970s, architects like Norman Foster, designed offices which simply covered their sites in a simple serviced void, right to the very edge, sealed only by a glass curtain-wall. In doing so they created the parameters for the contemporary office, highly serviced, computerised, essentially the same all across the world.
But the way people work is always changing. Open plan offices have periodically gone in and out of fashion, as have such innovations as the "bürolandschaft" and gimmicks such as "hot-desking". But bigger changes appear to be afoot due to developments in the globalised infrastructure of the internet. There is perennially much talk about home-working, but the prospect of people being productive away from the office is set off against the contempo-

照片提供：©Yuko Nagayama & Associates (Daici Ano)

神户的SISII花园式展示厅，日本，2010年
SISII in Kobe, Japan, 2010

当时，"第三工作地"成为流行的理念——无线网络的使用可以使工作几乎在任何地点都能完成，还存在一个关于办公室的想法，即尽管办公室普遍存在并具有机械灵活性，也可能产生更加灵活的类型，来适应未来工作实践的变化。仅举一例，在加利福尼亚，谷歌员工登上他们带有无线网络的巴士就被认为已经抵达工作地点。但是在未来的几年内其他灵活的类型又会发生什么变化？

设计和建筑将对社会变化起到重要作用，在一系列近期小型办公室的设计中，我们能看到可能的复合类型和工作空间，艺术家和设计者借鉴了零售店、画廊和家庭空间，以此来激发工作场所可能的新类型。例如，Savioz Fabrizzi建筑师事务所建造了一个小型绘画和雕刻工作室，这个工作室是他们在瑞士阿尔卑斯山地区的一个住宅翻新项目中的增建结构。它可以被看作一个简易"花园棚屋"的更大版本，但是这个工作室的外衣是光滑的混凝土，可能更能令人联想到城市中的艺术画廊。画廊作为复合型的工作空间，得以进一步的开发，由Haerynck Vanmeirhaeghe建筑师事务所为比利时画家Koen Van Den Broek建造。这里，在后工业化时期，办公室空间没有面向户外的窗口，设计师建造了居住室、画室、办公室和工作室，结合了基于传统19世纪绘画工作室的坚固性的美学，在某种程度上还有公司接待和展示区，也许期望成为21世纪成功的艺术家工作室。

Fraher建筑师事务所也研究了生活与工作空间的问题，这体现在该建筑事务所在英国伦敦为自己建造的办公室项目中。作为世界上极度缺乏空间的城市之一，他们的工作室占据了一个居住空间，从19世纪郊区的家穿过马路即是。这个工作室隐藏于砖墙后面，折叠方向复杂，且充分利用当地自然光，但是也降低了新开发项目的视觉冲击，它提供了这一设计公司工作的最小空间，避免了租借工作室的问题，但同时在工作进行的时候，也存在着没有扩建能力的问题。

Carlo Bagliani将自己之前的项目翻新成为一个建筑工作室，在这个翻新项目中展示了一种更加灵活的方法。作为意大利Sp10设计工作室的一部分，Bagliani在热那亚设计了一个地下车库，之后很快用作画廊空间。现在这里已经转变成一个小型建筑工作室，空间的整体昏暗被面向花园的长条窗户所带来的光明所抵消，项目中几乎不需要工作便可把画廊改造为工作室——一间会议室嵌入其中，档案室建在大厅的一侧。该

rary passion for "collaboration", Where the close proximity of other minds is considered the necessary link in the chain of creativity and innovation.

At the moment, "third place working" is becoming a popular idea – the availability of wireless internet allows for work to be done almost anywhere, and there is the thought that the office, despite its ubiquitousness and its mechanical flexibility, might have to become a much more flexible typology to accommodate changes in working practices into the future. To take just one example, in California, Google employees boarding their wifi-enabled buses are considered to have already arrived at work. But what other flexible typologies might occur over the coming years?

These are social changes in which design and architecture will have a significant role to play, and in a series of recent small office designs we can see possible hybrid typologies, work spaces for artists and designers which borrow from retail, gallery and domestic space to refresh the possibilities of what a workplace might be. For example, Savioz Fabrizzi Architectes have built a small painting and sculpture studio as an addition to a housing renovation they completed in the Swiss Alps. This may be seen as a larger version of the simple "garden shed", but it clothes itself in a sleek concrete idiom perhaps more reminiscent of an urban art gallery. The gallery as hybrid for workspace is explored further by the studios built for the Belgian painter Koen Van Den Broek by Haerynck Vanmeirhaeghe Architecten. Here, within a post-industrial space with no views to the outside, the designers have created a residence, painting studio, office and atelier, combining an aesthetic based upon the ruggedness of the traditional 19th century painter's studio, with a somewhat more corporate reception and display space, as perhaps is expected for a successful artist's studio in the 21st century.

The question of the live-work space is also investigated by the studio that Fraher Architects built for themselves in London, UK. In one of the most space-starved cities in the world, their studio occupies a residual space across the road from their own house in a 19th century suburb. Hidden behind a brick wall, and folded in awkward directions to take full advantage of the local light but also to minimize the visual impact of the new development, it provides a minimal space for this design practice to work from, avoiding on the one hand the problems of renting a studio, but at the same time existing without the capacity to grow as work comes in.

Carlo Bagliani's refurbishment of one of his own previous projects into an architecture studio shows a more flexible approach. As part of Italian design studio Sp10, Bagliani designed an underground garage in Genoa, which was put to use as a gallery space soon after. This has now been converted into a small architecture studio, the monolithic gloom of the space offset by the long strip-window's view out into the garden. Very little was needed to turn the gallery into the studio – a meeting room was inserted, and an

照片提供：©Haerynck Vanmeirhaeghe Architecten|Filip Dujardin

Koen Van Den Broek工作室，安特卫普，比利时，2013年
Koen Van Den Broek Studio in Antwerp, Belgium, 2013

项目证明了适合艺术展示的空间的空白和灵活性是如何被设计者所期望的。

类似于Bagliani的设计，位于稻田的Akitoshi Ukai/AUAU办公室在办公室与外部花园空间形成对话。但是在这个案例中，建筑师将单层建筑屋顶的户外空间与家具配套，在这种方法下，户外空间可以被认为是另外一处办公空间。这种方法模拟了当代办公大楼内扩大屋顶花园的做法，但是也期望成为所有人能在所有地方工作的一个地点，室内或室外，仅需开关他们的移动办公设备而达到此目的。

设计室和办公室能经常用作展示橱柜，向参观客户和合作者证明公司的价值。从这个意义上来说，办公室像一个店面，售卖想法而非物体。Yuko Nagayama联合事务所为日本神户的皮革品牌SISII设计的展示厅和办公室把这种逻辑发挥到极致，形成了具有讽刺"屋顶景观"概念意味的变体。三个独立的空间类型相互作用——展示厅的道路沿着一系列充满植物的空间迂回前行，展示的衣服挂在铁"树"形状的支架上。这些展示厅空间实际上建立在平台之上；而与地面同一水平并显然沉入地面里的是办公区，办公区的桌子看起来与参观者的脚在同一高度上。这些办公室里的工作人员也是展览的一部分，这是一个表面上看起来极端的条件，但是与之前见过的展示相似。

所有的工作室和工作空间展示了后现代工作空间与居住空间被重新发明的类型方式。由Estudio Elgue设计的巴拉圭亚松森Draft工作室与住宅增加了混合气质的设计，这种设计担心材料缺乏的未来，期待减少能源消耗。这座新建筑使用之前拆除的碎石建造而成，保留了砖块、木材，甚至电子设备。然而新建筑外部确实呈现了乡村的感觉，带有白灰砖和赤土色通风口，内部与其他工作室一样现代化，带有简单的调色砖、白色喷漆和玻璃。这座建筑不仅减少了浪费和能源，而且造价极低。

这些工作室都很小，没有能够容纳超过20个左右员工的能力。但是它们是结合工作和居住空间以及商业展览的良好范例。它们都回顾了现代办公室出现之前的时期，但是期望工作技术可能减小对大量单体建筑的需求，且可能会发展出更多生活和工作的复合空间的时期。

archive was built into one side of the large hall. It demonstrates how the blankness and flexibility that make a space suitable for the display of art is frequently what makes it suitable for the refined workplaces desired by designers.

Akitoshi Ukai/AUAU's Office in Rice Field, like Bagliani's design, brings the office into dialogue with an external garden space. But in this case the architect has provided an outdoor space on the roof of the single storey building which has been fitted out with furniture in such a way that it can be considered another office space. This mimics the increasing provision of roof gardens upon contemporary office block buildings, but also looks towards a point where anyone can work anywhere, inside or out, simply by switching on their mobile device.

Design studios and offices can frequently act as display cabinets, demonstrating the values of the company to visiting clients and collaborators. In this sense the office is like a shopfront, which sells ideas rather than objects. Yuko Nagayama & Associate's showroom and office for leatherware brand SISII, in Kobe, Japan, take this logic to an extreme, creating an ironic variation on the "bürolandschaft" concept. Three separate spatial types interact in the space – the showroom weaves its way around a series of spaces filled with natural plants, where clothing is displayed upon steel "tree" shaped hangers. These showroom spaces are actually set up on platforms – on floor level, thus apparently sunken into the ground, are the office spaces, whose desks appear at the same level of the visitor's feet. Here the staff of the office are almost part of the display themselves, a seemingly extreme condition, but one which is not too dissimilar to the displays seen before.

All of these studios and workspaces show ways in which the pre-modem typology of workplace-residences can be reinvented. Draft Studio and House designed by Estudio Elgue, in Asuncion, Paraguay, adds to the mix of a design ethos which looks forward to a future of material scarcity and reduced energy consumption. This new building is constructed from the detritus of a previous demolition, retaining the bricks, the timber, even the mechanical and electrical equipment. While the exterior of this new building certainly appears rustic, with its lime-washed brickwork and terracotta ventilation slots, the interior is as contemporary as any other of the other studios, with a simple palette of brick, white paint and glass. Not only does this construction cut down on waste and energy, but it also came in at an extremely low cost.

Each of these studios is small, and none of them are capable of accommodating a workforce of more than 20 or so people. But they are good examples of buildings which combine space for work and residence, as Well as space for commercial display. They both look back to a time before the modem office came into existence, but look forward to a time when the technology of work might reduce the need for massive single buildings, and more hybrid place of life and work might develop. Douglas Murphy

SISII花园式展示厅

Yuko Nagayama & Associates

这个展示厅与神户基町的一个办公室结合在一起,主要包含三个功能区:材料展示区、会客区和工作区。建筑师的意图是不按照分类来分隔空间,而是利用不同的高度来达到松散地连接和隔开不同空间的效果。连接结构的材质为大铁板。高架的铁板有加工过的锻孔,起到连接展示厅与办公室的作用。这些铁板下面种着各种植物,代表外面六甲山的生长环境,并且通过这些洞口揭示了自然之美。部分铁板被剥离或吊起,变成了会议区域或办公室里的大工作桌。随着铁板的延伸,不同的功能区应景而变,作为一个中性平台,与生机勃勃的植物相互辉映,是促进信息交流的绝妙场所。

SISII

This showroom combined with an office in Motomachi, Kobe is comprised of three actions which are customers watching the cloths, meeting people, and working. The architect's intention is not to categorically divide the space, but to loosely connect and disconnect with differences of heights. The connecting agent is a large iron sheet. A suspended iron sheet with wrought holes emulates the connection to showroom and office. Beneath these iron sheets, plants are grown to represent the outer Rokko Mountain's habitat and reveal the nature through these openings. The iron sheets are partially peeled and hoisted, and turned into a meeting space, and in the office into a large desk. Transforming roles accordingly to the spots, the iron sheet extends, as a neutral platform coexisting with the vital energy of the vegetations, and will enable an active site for informational exchange.

项目名称:SISII
地点:Kobe, Japan
建筑师:Yuko Nagayama
甲方:Au gré du vent Ltd.
设计表面积:144m²
竣工时间:2010.12
摄影师:©Daici Ano(courtesy of the architect)

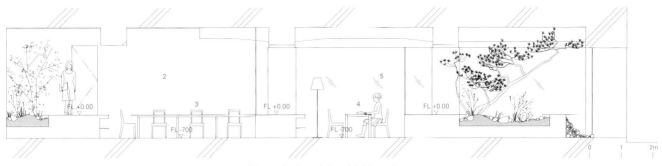

1 种植区 2 办公室 3 桌子 4 会议空间 5 展示厅
1. planting space 2. office 3. desk space 4. meeting space 5. showroom
A-A' 剖面图 section A-A'

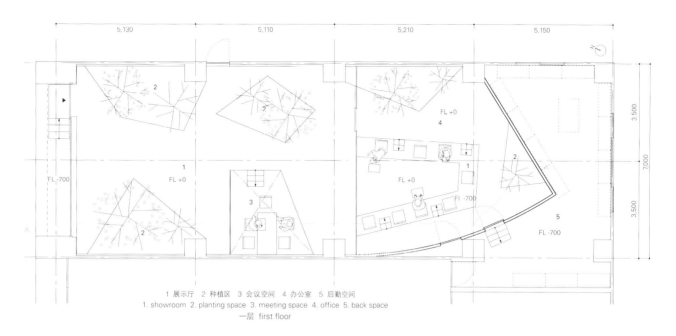

1 展示厅 2 种植区 3 会议空间 4 办公室 5 后勤空间
1. showroom 2. planting space 3. meeting space 4. office 5. back space
一层 first floor

Koen Van Den Broek工作室
Haerynck Vanmeirhaeghe Architecten

2008年，比利时画家Koen Van Den Broek在比利时莫克森沿安特卫普东部运河一处有点冷清的社区里购买了一个旧车库。他的作品在伦敦白立方画廊和科隆Figge von Rosen画廊的展览上获得了十分轰动的效应，他也因此获得国际认可，但在这之前的几年里，他在某种程度上已经厌倦了在一间狭小的工作室里作画。他需要新鲜的空气和充足的空间。他的画作尺寸不断地变化。为了追求从切实的远景来观赏自己的画作，他爱上了旧车库那长长的矩形空间以及远景效果。单一楼层的车库嵌入混杂的城市肌理中，汇集了储物空间、住宅、仓库以及全部扩建结构。砖块和疤痕、水泥地面，以及街道一侧的暗门都赋予了它强烈的材料特质。然而，由于车库本身没有正规的外立面，因此内部光线严重不足。

不过很快，在一位艺术评论家朋友沃特·戴维斯的提议下，他联系到了比利时根特市当地一位年轻的设计师Tijl Vanmeirhaeghe。两人对于艺术的直觉和兴趣立即联系到了一起。这是一次长久并富有成效的合作的开端，这间新工作室就这样诞生了。

Koen Van Den Broek用拍立得记录了自己横跨美洲之旅，其作品正是基于这些照片创作的。结果，这项工程也是源于设计师想要创造在不同星际间畅游体验的构思，成为一场在建筑学的道路上通往不同目的地的旅行。

建筑风格定位在一场由不安定的艺术家主导的永无止境的游戏，即他每一天对于新创意和新体验的探索和追求。

工作室的核心是一条长长的矩形走廊，汽车可以长驱直入来更换机油或零部件。空间是整个项目的主干，走廊两侧六扇高大的双开木门各自通往不同的功能区，构成了工作室的功能。地板平面中央部位与每扇门下的门槛之间的台阶增强了不同房间之间的过渡效果。如果将这间工作室比作宇宙空间站的话，这些门就相当于它的闸门。

左侧的前两扇门后是一间巨大的储藏室。而第三道门通向图书室，那里同时也是一间放映室。空间由桦木皮面板装饰，这是为了应John Lindman的要求而特别设计的。John Lindman是一位真正的桦木皮独木舟建造者，在美国东海岸斯波坎市的树林中生活和工作。因为这个房间并不用于绘画，设计师决定不在这个房间悬挂任何物品。白桦树皮的墙面既充当了覆层，本身又成了一幅画。

由于临近的房顶露台，工作室左翼无法自然采光。相反，工作室右翼则能够接收到阳光的照射，因此所有工作区都位于工作室的这一部分。面向图书室的门是通向一处面积为9.8m×9.8m的"白立方"空间的入口。Koen的大多数画作都是在这里创作出来的。

同时，这个房间的重要性还在于将尝试用于举办不同的展览。这是工作室空间背景下的一间微型博物馆。工作室中央部位右侧的中门通向一间绘画工作室，那里现在被用作了办公间。一面北向的巨大玻璃立面朝向一个天井。天井有独立的门能够通到中央空间，但它更主要的是充当了一个隔断，将工作空间与面向街道一侧的小歇脚处分隔开来。一间小浴室、厨房和卧室组成一个基本的居住区，与艺术家的工作室分隔开来。

Koen Van Den Broek Studio

In 2008 the Belgian painter Koen Van Den Broek bought an old garage in the somewhat inhospitable neighbourhood of Merksem along the canal at the eastside of Antwerp, Belgium. In the years before he received international recognition for his work with sensational exhibitions in the White Cube, London and Figge von Rosen Gallery, Cologne. In a way he was fed up with painting in a small studio. He needed air, and space. His paintings were changing size. In search for a literally distant view on his own work, he fell in love with the old garage with its long rectangular spaces, its perspectives. The garage had a single floor embedded in a mingled urban fabric of storages, houses, warehouses and extensions of all kinds. It had a strong material identity by its bricks and scars, its concrete pavement, its blind gates at the streetside. There was a critical lack of daylight as the garage has no proper facades.

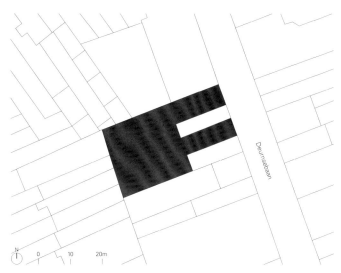

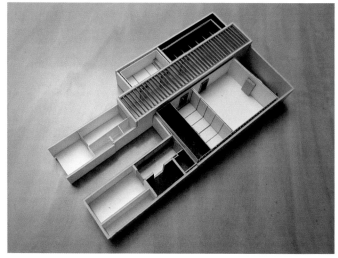

东北立面 north-east elevation

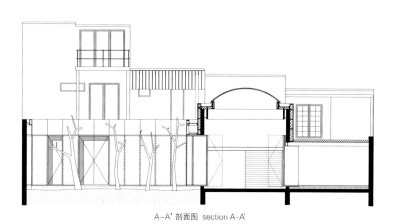

A-A' 剖面图 section A-A'

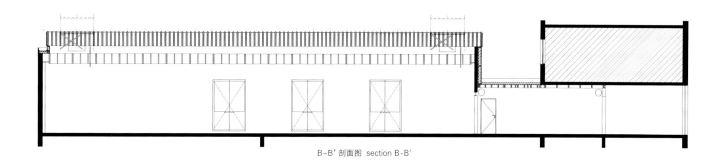

B-B' 剖面图 section B-B'

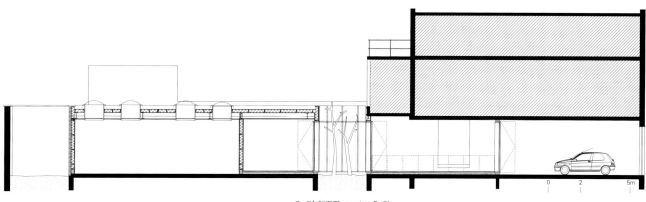

C-C' 剖面图 section C-C'

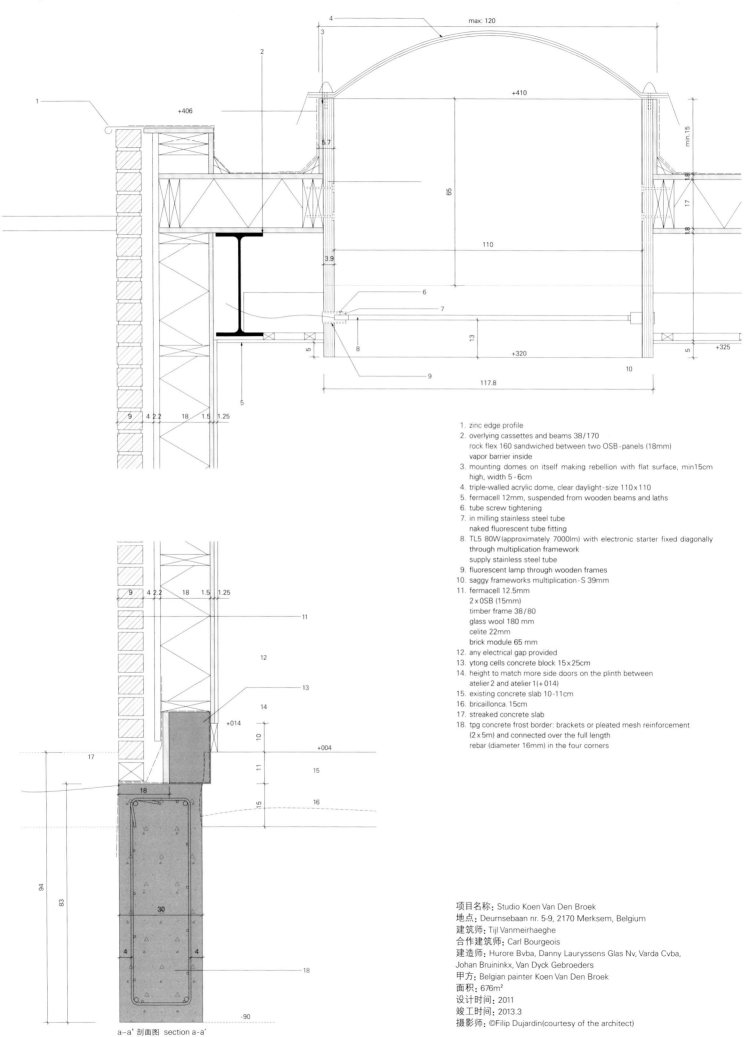

a-a' 剖面图 section a-a'

1. zinc edge profile
2. overlying cassettes and beams 38/170
   rock flex 160 sandwiched between two OSB-panels (18mm)
   vapor barrier inside
3. mounting domes on itself making rebellion with flat surface, min15cm
   high, width 5 - 6cm
4. triple-walled acrylic dome, clear daylight-size 110 x 110
5. fermacell 12mm, suspended from wooden beams and laths
6. tube screw tightening
7. in milling stainless steel tube
   naked fluorescent tube fitting
8. TL5 80W(approximately 7000lm) with electronic starter fixed diagonally
   through multiplication framework
   supply stainless steel tube
9. fluorescent lamp through wooden frames
10. saggy frameworks multiplication-S 39mm
11. fermacell 12.5mm
    2 x OSB (15mm)
    timber frame 38/80
    glass wool 180 mm
    celite 22mm
    brick module 65 mm
12. any electrical gap provided
13. ytong cells concrete block 15 x 25cm
14. height to match more side doors on the plinth between
    atelier 2 and atelier 1(+ 014)
15. existing concrete slab 10-11cm
16. bricaillonca. 15cm
17. streaked concrete slab
18. tpg concrete frost border: brackets or pleated mesh reinforcement
    (2 x 5m) and connected over the full length
    rebar (diameter 16mm) in the four corners

项目名称：Studio Koen Van Den Broek
地点：Deurnsebaan nr. 5-9, 2170 Merksem, Belgium
建筑师：Tijl Vanmeirhaeghe
合作建筑师：Carl Bourgeois
建造师：Hurore Bvba, Danny Lauryssens Glas Nv, Varda Cvba,
Johan Bruininkx, Van Dyck Gebroeders
甲方：Belgian painter Koen Van Den Broek
面积：676m²
设计时间：2011
竣工时间：2013.3
摄影师：©Filip Dujardin(courtesy of the architect)

Shortly after, and on recommendation of his friend, the art critic Wouter Davidts, he contacted the young Ghent based architect Tijl Vanmeirhaeghe. There was an immediate connection between their artistic instincts and interests. It was the beginning of a long and fruitful collaboration which ultimately resulted in the birth of the new studio.

The work of Koen Van Den Broek is based upon polaroid pictures that document his travels throughout America. As a result, this project was conceived by the architect as a travel through different universes, an architectural roadtrip with various destinations. The architecture is directed as an endless play dominated by the restlessness of the artist, his everyday search for new inventions and experiences.

At the core of the studio there's a long and rectangular atelier where once cars run in and out in order to change oil or parts. The space serves as the backbone of the project, with six double doors connecting it to six totally different spaces which constitute the program of the studio. A little rung between the floor level of the central space and the sill of the different doors strengthens the effect of transition when changing rooms. If the studio could be compared to a space station, the doors are its sluice-gates.

The first couple of doors on the left both serve a big storage space. The third door leads to the library which is also a projection room. This space is cladded with birch bark panels which were specially made on demand by John Lindman, an authentic birch bark canoe builder living and working in Spokane, in the midst of the forests of the American east coast. As this room is not a space for drawing or painting, the architect decided to exclude every need or possibility to hang up things. The birch bark is wall, cladding and painting on its own.

The left wing of the studio could not be illuminated naturally because of an upstairs roof terrace of the neighbour. On the contrary in the right wing there was opportunity to take daylight, so all the working spaces are located in this part of the studio. Facing the doors to the library, there's the entrance to a square "white cube" of 32 by 32 feet. Here the majority of Koen's paintings take place. But the room is also crucial for setups for different exhibitions which can be tried out here. It's a micro museum in the context of a studio space. The middle doors at the right of the central atelier lead to a drawing studio which nowadays serves as an office space. It has a big north oriented glass facade facing a mineral patio. This mirror patio has its own doors to the central space, but mostly functions as a gap between working spaces and a small pied-à-terre facing the street side. A small bathroom, kitchen and bedroom constitute an essential dwelling, cut loose from the artist's studio.

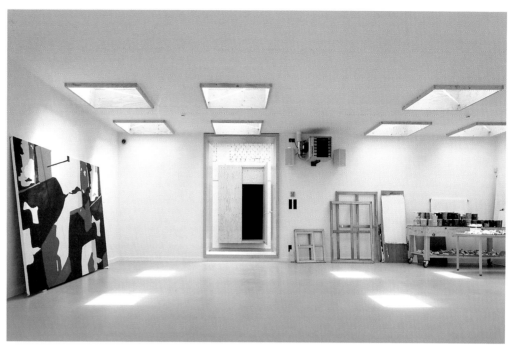

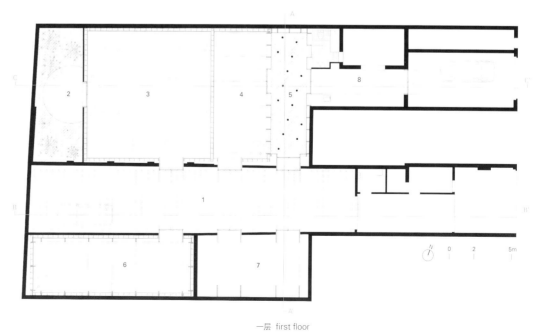

1 工作室
2 棕榈树花园
3 白立方
4 绘画工作室
5 露台
6 图书室
7 储藏室
8 住宅

1. atelier
2. palm garden
3. white cube
4. drawing office
5. patio
6. library
7. storage
8. residence

一层 first floor

露台立面图示
patio elevation figure

| A立面图 | B立面图 | C立面图 | D立面图 |
| elevation A | elevation B | elevation C | elevation D |

起居与工作相融合 Live/Work Hybrids

Draft工作室与住宅
Estudio Elgue

在我们所生存的世界里，一切在用完之后都可以被丢弃。几乎所有东西都在被快速地消耗和浪费，然后再循环往复地被消耗和抛弃。而建筑消耗了地球上65%的自然资源和40%的能源，产生的剩余物占地球剩余物总量的50%。生产水泥、钢铁、石灰、陶瓷、玻璃、铝等等的工业加热用炉所需的能量有所增加，再加上把这些材料从一个洲运送到另一个洲所需要的能源，能源最终全都消耗在建筑上。

建筑师提出，在探索减轻坏境破坏所带来的影响的可能性的同时，对废弃材料的再利用或复原既是个人的挑战，也是大家共同的挑战。于是，魔力产生了，那些被忽视的材料成为建筑中的主角，从死亡中复活，熠熠生辉，重写了一部生机勃勃的、满载尊荣与乐观的建筑史。

这一委托任务包含一个工作和生活的混合空间：家和工作室，每一部分都可以单独使用，或者作为一个单一的部分，一个城市生产单元。

原有建筑部分（砖、木料、瓷砖、瓦片、通风设施、电力设施、管道设施等等）完全被拆除和破坏，只剩下一些残垣断壁。项目正是通过恢复和重新使用这些残余材料建成的。例如，砖制的通风孔都是不加选择地使用这些修复的材料。因此你可以清晰地看到墙壁上那些徒手画作的粗犷线条。为了保持这样的质地，建筑师有意放弃了"精细的成品"。

大部分现存的墙壁由灰泥和颜料构成，而该项目的墙壁则是带气孔的新表面。建筑师采用符合生物气候学的建筑标准，在面向盛行风方向将带气孔的空心砖嵌入墙壁或板材内，以用于自然通风，减少空调的使用。气孔的光影在地板上投射出马赛克图案，正如流逝的光阴在时空中所留下的足迹。

建筑采用自然排风系统来不断地更新室内空气。较低的窗户用来吸收新鲜空气，较高的窗户则可以排出热空气。同时可伸缩的光传播面板可使室内嵌板和屋顶漫射较平时两倍的光线，减少了日间的人工照明。

修复的钢化玻璃无法切割，嵌入了地板和墙壁中，排列成所需样式。木制品则用来盛装垃圾，以及用作托盘。同时，建筑中拆毁的大量灯具也得到再利用。

项目占地140m²，总投资13000美元，不到巴拉圭同等传统建筑投资额的二分之一。建筑在形象上具有双重含义：就其密闭而言，建筑在日间吸收了周围植被的树阴，像是随时间而改变的投影屏幕；而日晚上在光线的作用下，晃动的墙体表皮和气孔投射出阴影，就像是一幅素描画。

Draft Studio and House

We live in a world where everything is disposable and almost everything is fast consumption and waste, and then returns to consume and discard, where the construction is liable to spend about 65% of natural resources, 40% of the energy and generate 50% of the total residues of the planet. The amount of energy required for heating industrial furnaces to produce cement, steel, lime, ceramic, glass, aluminum, etc. is added, plus that required to transport materials from one continent to another and finally, the energy is consumed in buildings.

The architects propose an alternative reuse of waste materials and recovery, as individual and collective challenge, as people are exploring the possibilities to mitigate that impact of environmental damage described.

A kind of alchemy takes place then, to turn that ignored materials, the main protagonist of the work, resurrected from the dead, reborn so vigorously in complicity with light to rewrite a story full of dignity and optimism.

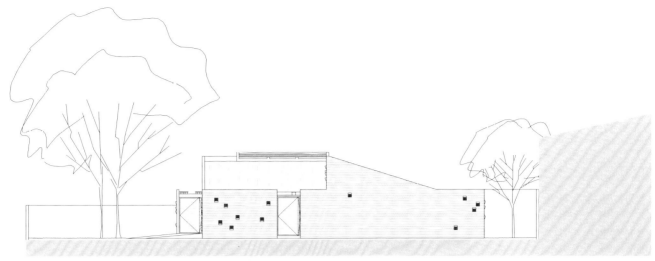

西立面 west elevation

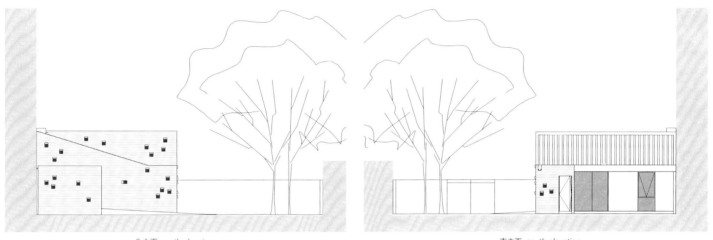

北立面 north elevation　　　　　　　　　　　南立面 south elevation

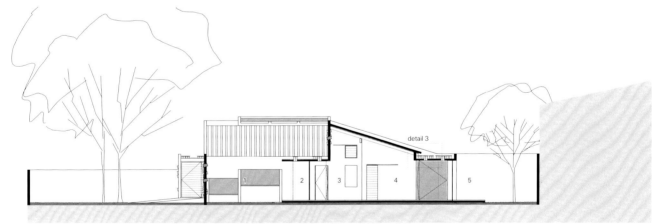

1 室内花园　2 入口长廊　3 餐厅　4 起居室　5 花园
1. internal garden 2. entrance gallery 3. dining room 4. living room 5. garden
A-A' 剖面图　section A-A'

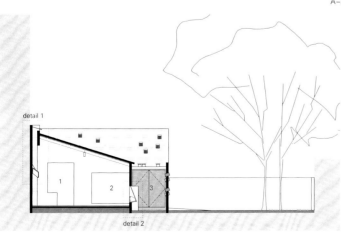

1 图书馆　2 办公室　3 室内花园
1. library 2. office 3. internal garden
B-B' 剖面图　section B-B'

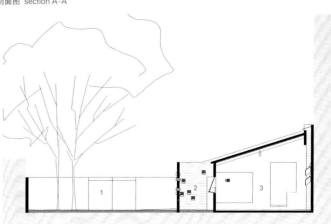

1 停车场　2 室内花园　3 工作区
1. parking 2. internal garden 3. work area
C-C' 剖面图　section C-C'

项目名称：Boceto
地点：Teniente Héctor Vera Street 1.666 and San Martin Avenue, Asunción, Paraguay
建筑师：Luis Alberto Elgue Sandoval, Cynthia Solis Patri
项目领导：Cecilia Román
用地面积：276m²
总建筑面积：140m²
设计时间：2012.7
施工时间：2013.4—10
摄影师：courtesy of the architect-p.86, p.91
©Leonardo Finotti-p.84~85, p.88, p.90, p.93, p.94~95

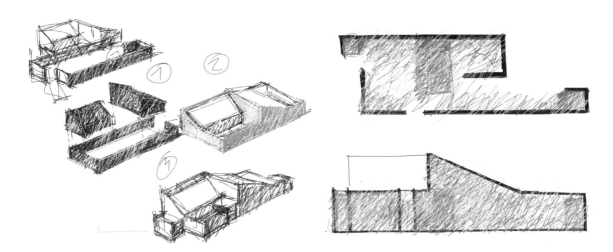

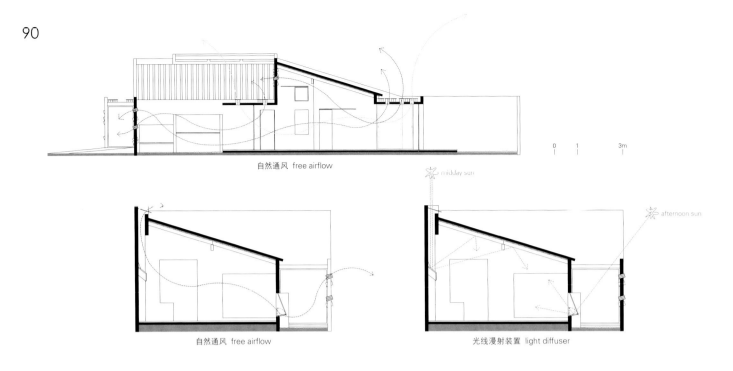

自然通风 free airflow

自然通风 free airflow   光线漫射装置 light diffuser

The commission consists of a hybrid space: home and workshop room, for work life and livelihood, where each part can operate independently or as a single piece, an urban productive unit.

The project arises from the complete dismantling and demolition of an existing building(bricks, timber, tiles, shingles, vents, electrical and plumbing fixtures, etc.), leaving some walls standing and using these materials fully recovered in new developments. In the case of the brick, implicitly assuming the porosity using these recovered materials happens without selection. So you can clearly see the thick lines of freehand drawings. In order to keep the texture, the architects intentionally gave up "good ending".

Most of the existing walls are made of plaster and color, however, the walls of this project have a new porous skin. Rehearsed criteria-bioclimatic architecture proposes breathing pores in hollow common brick, embedded in walls and slabs for natural air circulation, facing the prevailing winds, reducing the use of air conditioner. These pores also projected light mosaics on the floor, as the day progresses, marking the pass of time in the space.

Using the low windows for fresh air and the upper ones to expel the heat air, the architects renovated the air constantly with a natural extraction system. They also propose a retractable light diffusing panels for double daylight to rebound plates and roof which reduced the use of artificial illumination during the day.

Cut tempered glass is unable to be recovered, so the architects embedded it in floors and walls, adjusting as needed. Woodwork was used for packaging waste, and pallet type. The vast majority of the lighting fixtures were reused from the demolished building. The total cost of the project is $13,000 for 140m² which is much less than half of the cost of a traditional construction in Paraguay. The building has a double meaning in its image; by its hermetic condition, by day it absorbs surrounding vegetation by shade, as a projection screen that mutates along hours; by night under the influence of light, vibrating skin and pores cast shadows as if it were a sketch. Estudio Elgue

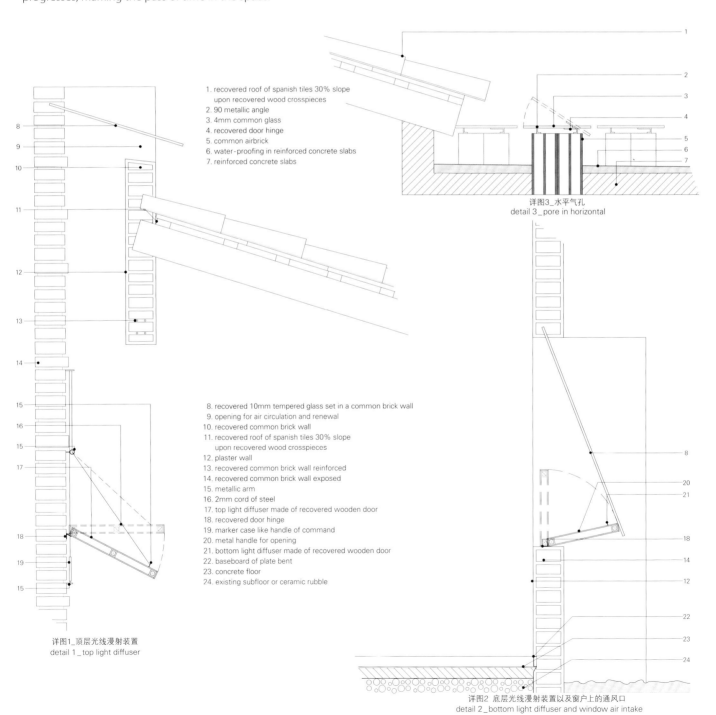

1. recovered roof of spanish tiles 30% slope upon recovered wood crosspieces
2. 90 metallic angle
3. 4mm common glass
4. recovered door hinge
5. common airbrick
6. water-proofing in reinforced concrete slabs
7. reinforced concrete slabs

8. recovered 10mm tempered glass set in a common brick wall
9. opening for air circulation and renewal
10. recovered common brick wall
11. recovered roof of spanish tiles 30% slope upon recovered wood crosspieces
12. plaster wall
13. recovered common brick wall reinforced
14. recovered common brick wall exposed
15. metallic arm
16. 2mm cord of steel
17. top light diffuser made of recovered wooden door
18. recovered door hinge
19. marker case like handle of command
20. metal handle for opening
21. bottom light diffuser made of recovered wooden door
22. baseboard of plate bent
23. concrete floor
24. existing subfloor or ceramic rubble

详图1_顶层光线漫射装置
detail 1_top light diffuser

详图3_水平气孔
detail 3_pore in horizontal

详图2_底层光线漫射装置以及窗户上的通风口
detail 2_bottom light diffuser and window air intake

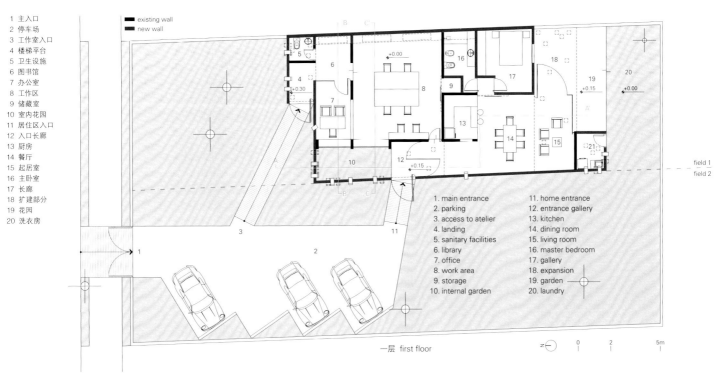

1 主入口	1. main entrance
2 停车场	2. parking
3 工作室入口	3. access to atelier
4 楼梯平台	4. landing
5 卫生设施	5. sanitary facilities
6 图书馆	6. library
7 办公室	7. office
8 工作区	8. work area
9 储藏室	9. storage
10 室内花园	10. internal garden
11 居住区入口	11. home entrance
12 入口长廊	12. entrance gallery
13 厨房	13. kitchen
14 餐厅	14. dining room
15 起居室	15. living room
16 主卧室	16. master bedroom
17 长廊	17. gallery
18 扩建部分	18. expansion
19 花园	19. garden
20 洗衣房	20. laundry

一层 first floor

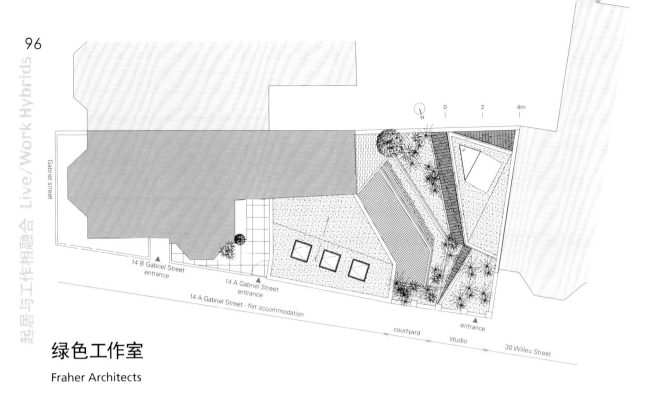

绿色工作室
Fraher Architects

绿色工作室坐落于蝴蝶楼对面,是花园式家庭工作空间,展示了极富创意的建筑实践。该工作室位于伦敦的东南部,是应各层主管们的需求而建,目的是平衡年轻人家庭与日益增长的工作负担之间的关系。

工作室的外形和方向取决于详细的日照分析,以期尽量减少其对周围建筑物的影响,同时确保对其中的花园和工作空间的最大程度的自然采光。

上下分层的接地形式有助于隐藏占地面积,并且使花园和工作室之间的地面景观过渡自然而生动。包在不锈钢丝网里的梯田种植床和野生花卉覆盖的屋顶使立面看起来清新自然,替代了我们失去的生活环境。

精选的高性能玻璃结合超级绝缘材料和强大的自然通风措施,保证了该建筑无需加热或冷却系统。巨大的太阳能电池板和热存储设置可以为厨房和淋浴室提供热水。

项目名称:The Green Studio
地点:London, UK
建筑师:Fraher Architects Ltd
结构工程师:Constant Structural Design
木工:Fraher and Co Ltd
有效楼层面积:32m²
施工时间:2013.3—10
摄影师:©Jack Hobhouse(courtesy of the architect)

The Green Studio

Sited opposite the Butterfly House, the studio is a garden based creative home work space for architectural practice. Situated in southeast of London the building was driven by the directors' need to balance a young family with an increasing workload.

A-A' 剖面图 section A-A'

The studio's shape and orientation have resulted from a detailed sunlight analysis minimizing its impact on the surrounding buildings and ensuring high levels of daylighting to the garden and work spaces.

The split levels and grounded form help to conceal its mass and facilitate the flowing ground-scape transition between the garden and studio. Clad in a stainless steel mesh the terraced planter beds and green wild-flower roofs will combine to green the facade replacing the lost habitat.

Carefully orientated high performance glazing combined with super insulation and a robust natural ventilation strategy means the building requires no heating or cooling. Hot water for the kitchen and shower is provided by a large solar array and thermal store.

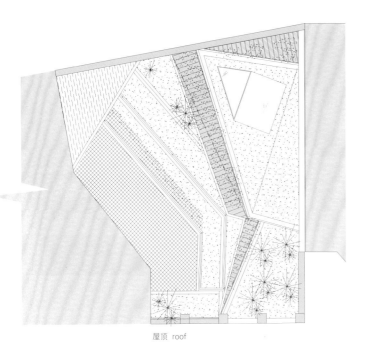

屋顶 roof

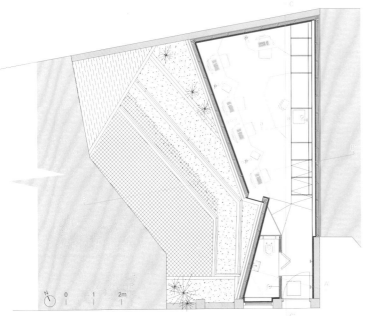

一层 first floor

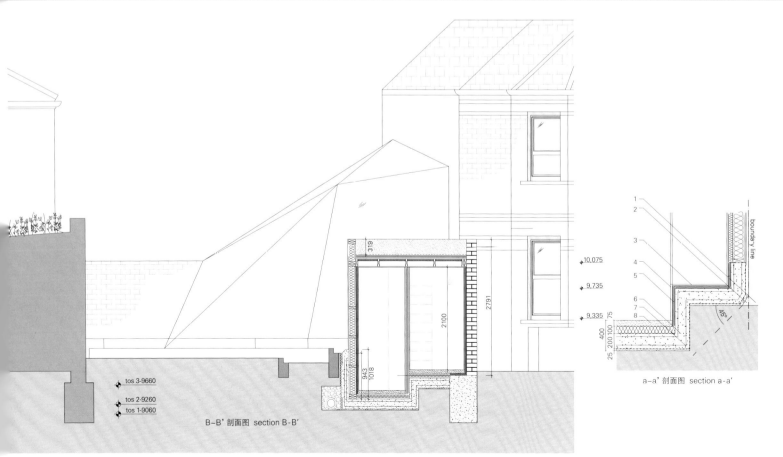

B-B' 剖面图 section B-B'

a-a' 剖面图 section a-a'

solar thermal array
green roof
(doc lindum)
double glazed
fixed panel
double glazed
roof light
stainless metal
mesh

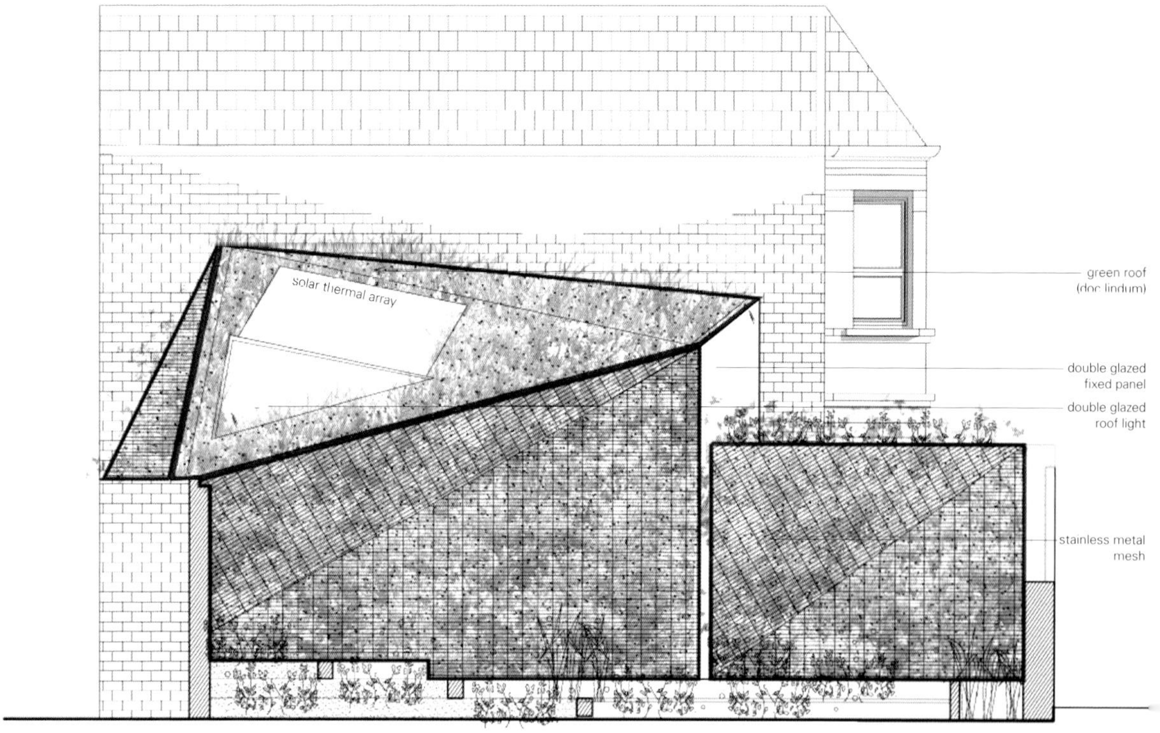

立面_窗户和室外材料
elevation_windows and external materials

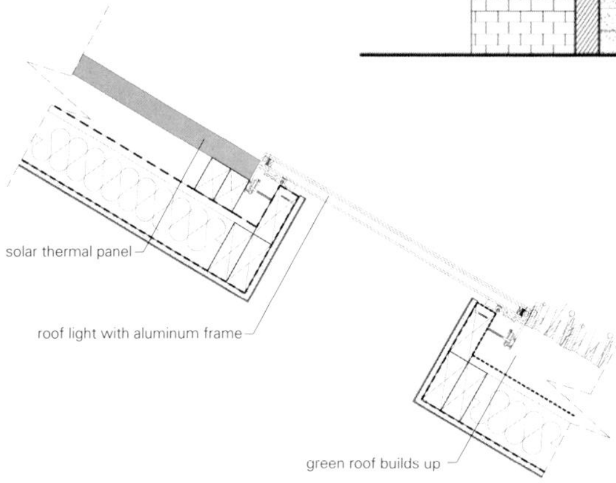

solar thermal panel
roof light with aluminum frame
green roof builds up

b-b' 剖面图 section b-b'

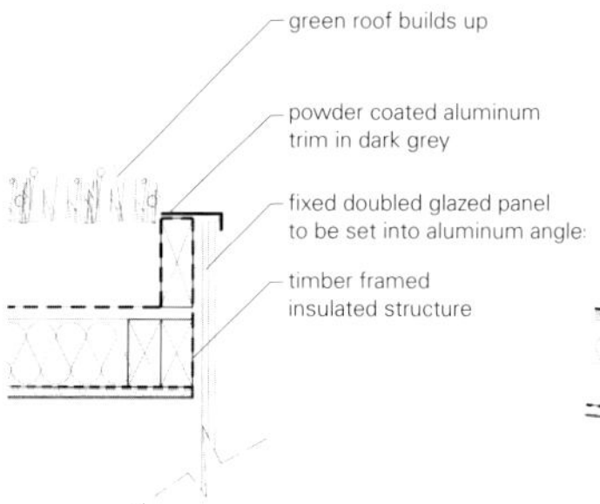

green roof builds up
powder coated aluminum
trim in dark grey
fixed doubled glazed panel
to be set into aluminum angle
timber framed
insulated structure

c-c' 剖面图 section c-c'

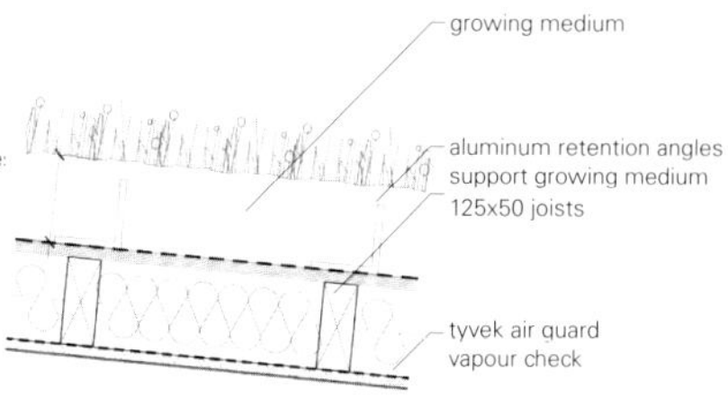

growing medium
aluminum retention angles
support growing medium
125x50 joists
tyvek air guard
vapour check

d-d' 剖面图 section d-d'

1. base of neighbours' foundation
2. filler
3. riw flexiseal
4. 200mm rc-slab
5. 100mm rigided insulation
6. dpm 1200g polythene
7. 25mm sand blinding
8. 75mm fine polished concrete screed
9. existing boundary wall
10. existing foundation

10
9
8
5
3
4
7
10,050
10,075
9,735
9,335
9,135
75
100
200
25
400

详图1 detail 1

2200
255
10,000
tos 3-9660
tos 2-9260
tos 1-9060
detail 1
0
2
5m

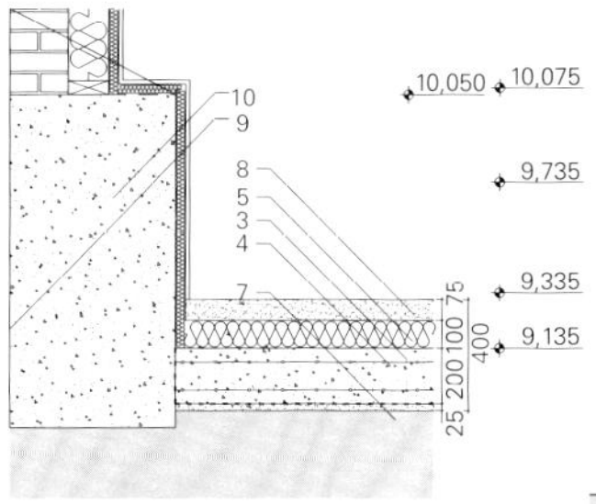

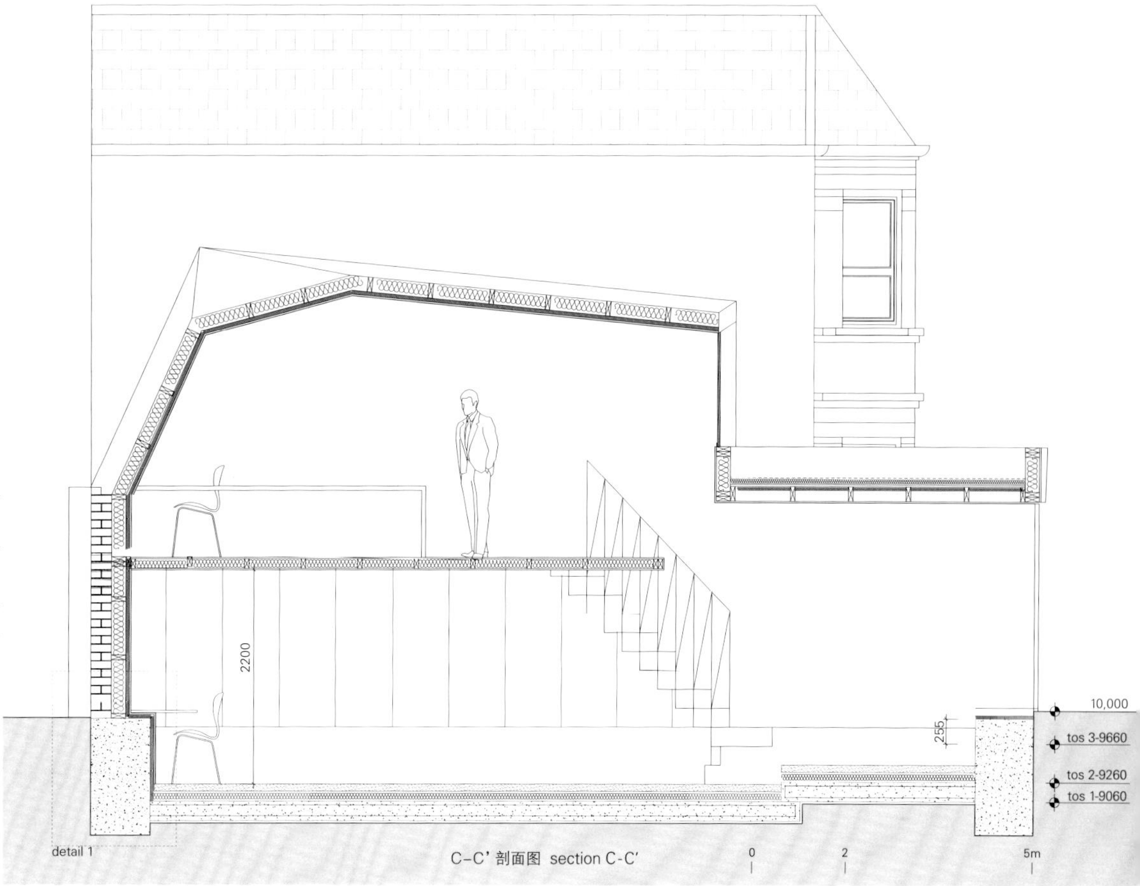

C-C' 剖面图 section C-C'

地下车库建筑工作室
Carlo Bagliani

当设计师自己就是自己的客户时,任务就会更简单一点,因为建筑师们允许自己免受委托合同的限制。这个项目的目的是利用合理的预算将一间停车场改建成为自己的一间建筑工作室。

建筑师只需要为设计师和一些设施准备很多工作间。

这片区域之前是一块旧菜园,在六层的相邻建筑于20世纪60年代时幸存了下来。

事务所在2005年与Sp10建筑事务所合作共同设计了这个停车场。

原有建筑最初被设计为一个停车场,是因为按照当时的规定,这处地下空间只允许建地下车库,但当时建筑师们已经考虑过将它用作艺术或其他类似的空间。所以这座被称为"菜园中的艺术空间"的建筑已在2011年刊登于意大利最重要的建筑设计杂志《Casabella》上。

这处空间非常不错,由混凝土制成,非常坚硬牢固。设计师们很喜欢这里。

因此,在重建项目的设计过程中,在某种程度上保留原有空间的特性也是很重要的。

这处空间还曾被一位抽象表现派画家朋友Federico Palerma用作画室使用了仅仅一年时间。他在冬天曾使用火炉取暖。因此为了纪念这部分故事,设计师决定保留外墙上这块烟熏的黑点(你可以在带有两把Toshiuki Kita设计的单人沙发的照片中看到它)。

像这样,过去常常影响到日后的生活也是一件非常美好的事。

最重要的是要实现一个有序的空间,让人们可以停下来反思生活。

为尽量尊重原有开放空间的特性,设计师还决定只引入三处基本的分隔区域:档案馆,像是主要开放空间里的一块安静低矮的巨石,另外隔出了小厨房和放置绘图仪的地方;然后是会议室和卫生间。这块巨石有着非常重要的作用,它是这处开放空间里的一个生动的存在,你可以站在它周围的不同角度来获得不同的共鸣,这也许跟风水文化有一定的关系。

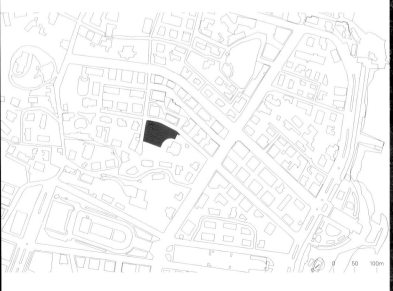

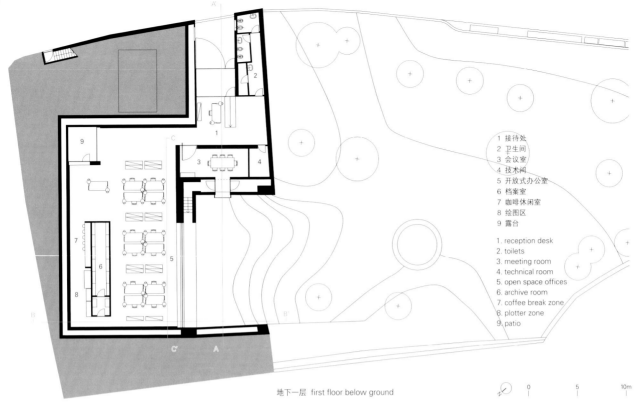

	接待处
	卫生间
	会议室
	技术间
	开放式办公室
	档案室
	咖啡休闲室
	绘图区
	露台

1. reception desk
2. toilets
3. meeting room
4. technical room
5. open space offices
6. archive room
7. coffee break zone
8. plotter zone
9. patio

地下一层 first floor below ground

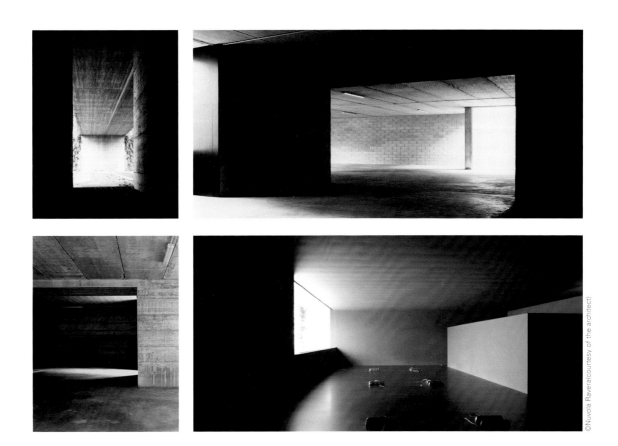

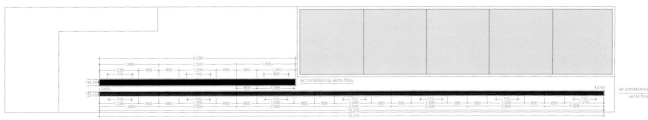

C-C'立面图_空调口 elevation C-C'_air conditioning vents

建筑师也尽量将电气系统和空调装置对天花板和墙壁的污染降到最低,因为整洁有助于精神的放松与专注。

建筑师认为这处空间像一座面向现代僧侣开放的古老的图书馆,带有金属的桌子,每个桌上配有桌灯,一个紧挨着一个。

因为预算非常有限,建筑师只好使用石膏板作为天花板和墙壁材料,用天然橡胶作为地面材料。所有事物都是黑色的,总是有助于注意力的集中。它肯定非常舒服。

白色金属家具是事务所与Sp10建筑事务所合作时由Antonio Norero设计的;它们原本是为一项名为"食物与设计"的展览设计的,当时事务所受委托为展览设计开场。所以对这些家具而言,这也算是二次利用,意味着第二次的生命!

Underground Garage Architecture Office

The assignment was a little easier than usual because the architects were their client, so they were allowed to work being free from the obligation of the client. The purpose was to convert a parking garage in an office of architecture for their office, using reasonable budget.

The architects just needed about a dozen of workspaces for designers with some facilities.

The area in origin was an old vegetable garden, a survivor in a neighborhood made of six stories building realized in the sixties. In 2005, the office designed the parking garage together with Sp10, the partner architectural office.

This original building was designed first as a parking garage be-

项目名称:Office of Architecture in an Underground Garage in Genova
地点:Genova, Italy
建筑师:Carlo Bagliani
合作建筑师:Stefano Mattioni, Pamela Cassisa
用地面积:5,800m²
总建筑面积:448m²
有效楼层面积:340m²
设计时间:2012
施工时间:2013
竣工时间:underground garage _ 2007, refurbishment as office of architecture _ 2013
摄影师:©Anna Positano(courtesy of the architect)(except as noted)

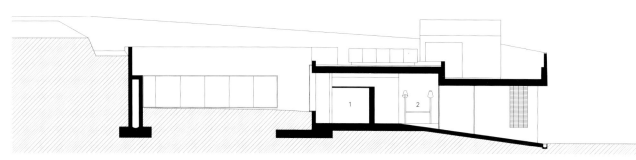

1 会议室 2 接待处 1. meeting room 2. reception desk
A-A' 剖面图 section A-A'

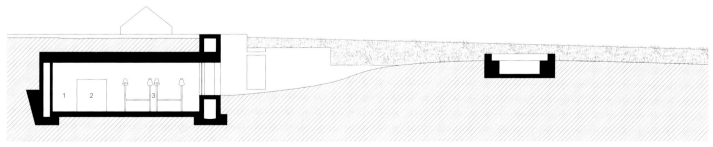

1 绘图区 2 咖啡休闲室 3 档案室 1. plotter zone 2. coffee break zone 3. archive room
B-B' 剖面图 section B-B'

cause of the legislation at the time that didn't permit to build something else, but the architects were already thinking at a space for art or something similar. So the building described as "a space for art in a vegetable garden" was already published in 2011 by the most important Italian architecture magazine, *Casabella*.
It was a very nice space, made in concrete, very hard, tough. The architects loved it.
So designing the refurbishment, it was also important to save something of the character of the original space.
For just one year the space was also used as atelier by a friend, a kind of abstract expressionist painter, Federico Palerma. He used a stove to heat during winter, so to remember this part of the story the architects decided to leave the black spot of smoke on the wall outside (you can see it in the shot with the two black armchairs by Toshiuki Kita).
So it's nice that very often the past takes its influence on what comes later.
The most important thing was to realize an orderly place in order to stop and meditate.
The architects also tried to respect the character of the original open space, so they choose to introduce only 3 essential separated areas: the archive, like a quiet low-rise monolith in the main open space, a space that also separates the area for the kitchenette and the plotters; then the meeting room and the restrooms. The monolith is very important, it's an alive presence in the open space, you can stand around it (him) on the different sides, and get different vibes, perhaps it has a relationship with fengshui culture.
Also about the electrical system and air conditioning the architects tried to disturb the minimum of the cleanness of the ceiling and of the walls, because this neatness can contribute to the relaxation and concentration of the mind.
The architects think that this space runs like an old library for contemporary monks, with their metallic desks with their own lamps, one behind the other.
The architects had a small budget so they just used plasterboard for ceiling and walls and natural rubber for the floor. Everything is always black to contribute to concentration. It was a bet and yes it's really very comfortable.
The white metallic furniture was designed by Antonio Norero when the architects were partners in Sp10; in origin it was designed for an exhibit for which the office had the assignment for designing the setup, and the title was "Food and Design". So for the furniture it's also a kind of reutilization, a second life!

Carlo Bagliani

稻田中的办公室
Akitoshi Ukai / AUAU

位于稻田中的DEN-EN办公室是一家劳务派遣公司的总部。

该公司拥有众多正在进行的项目,所以信息沟通是非常重要的。工作人员也应有明亮的空间可用。

该区域被设计为舒适且非正式的工作空间。除了室内,室外也提供了户外办公环境。以下设计元素可以确保工作人员舒适地四处走动,并且所看之处皆为风景。

临时员工和工作人员可以在平台或较高的会议区交谈。白板上的文字和图形都可以与其他人员共享。楼梯也可以成为讨论的地方。

平台的另一侧有一片稻田。脑力工作者可以向外望去,欣赏这片宁静的风景。这样的办公室可以激发创造力,且提高工作效率。办公室中央的厨房里设有咖啡机。工作人员可以在屋顶享受午餐。

为了营造一个没有隔间的大房间,设计中采用了层压的柏木作为横梁。这些木材产自爱知县,还用于建筑外墙,具有自然、亲切的特征。建筑师在地面上采用了柱和梁的结合设计,然后把它们用起重机吊起并垂直放置在地基的上面。

剖面设计中充分考虑到通风路径,并与郊区气候、景观相结合,同时引进屋顶绿化以解决环境问题。

会议区屋顶上设有一个天篷。在讨论的间隙,与会者可以抬头看到透过天篷射来的柔和的光线。这个天篷也具有烟囱的功能,可以加速从稻田吹来的风,所以即使在夏天也可以感觉到习习凉风。

Office in Rice Field

This office, in the middle of a rice field, is the headquarters of a staffing company.
The company has many ongoing projects. Communication is extremely important. Workers have well lit workspaces available to them.
The area is designed to be a comfortable and informal working space. This office has an outdoor environment as well as an interior. The following design elements meant to ensure that the workers are comfortable while walking around. There are vistas wherever they look.

项目名称:DEN-EN Office
地点:Japan
建筑师:Akitoshi Ukai/AUAU
结构工程师:Masaichi Taguchi/TAPS
建筑工程师:Watanabe Construction Company
室内设计师:Akitoshi Ukai/AUAU
景观设计师:Akitoshi Ukai/AUAU
甲方:MIYAKO Co., Ltd.
用地面积:572.9m² / 总建筑面积:228.6m²
有效楼层面积:228.6m²
总建筑规模:one story above ground
结构:wood / 室外饰面材料:wood / 室内饰面材料:plaster
竣工时间:2012.4
摄影师:©Masaya Yoshimura (courtesy of the architect)

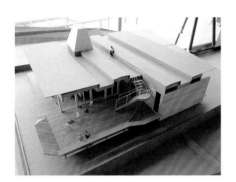

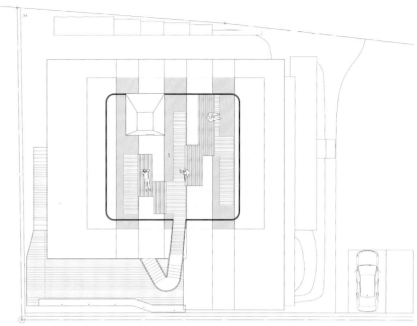

1 屋顶花园办公室 1. roof garden office
屋顶 roof

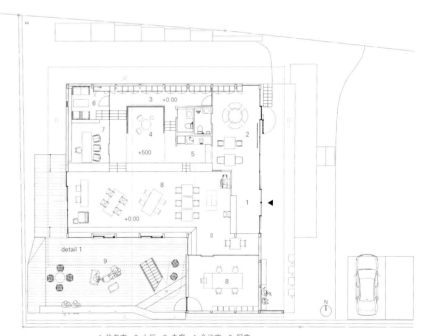

1 休息室 2 大厅 3 走廊 4 会议室 5 厨房
6 接待室 7 操作室 8 办公室 9 木质平台办公室
1. lounge 2. lobby 3. corridor 4. meeting room 5. kitchen
6. reception room 7. operation room 8. office 9. wood deck office
一层 first floor

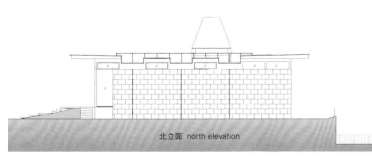

北立面 north elevation

西立面 west elevation

Temporary staff and workers can talk to each other on the deck or in a raised meeting space. Characters and graphics on the white board are to be shared by other workers. Stairways become places for discussion.

There is a rice field on the other side of the deck. Intellectual workers can look out at a peaceful landscape. The office is a place for creativity and efficiency. There is a coffeemaker in the kitchen in the middle of the office. Some workers have lunch on the roof.

To create a large room without partitions. laminated cypress was used as beams. The cypress grew locally in Aichi was used for the exterior walls and enhances the nature-friendly design. The architects arranged a combination of columns and beams on the ground. They were then lifted with a crane and placed upright on top of the foundation.

A sectional ventilation path was designed based on the suburban climate and landscape. Roof vegetation was introduced to solve environmental problems.

There is a canopy on top of the meeting space. When a discussion stalls, participants can look up at the soft light in the canopy. The canopy is also a chimney. Wind from the rice fields, is accelerated by the chimney. The wind blows even in the summer.

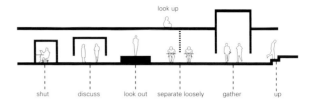

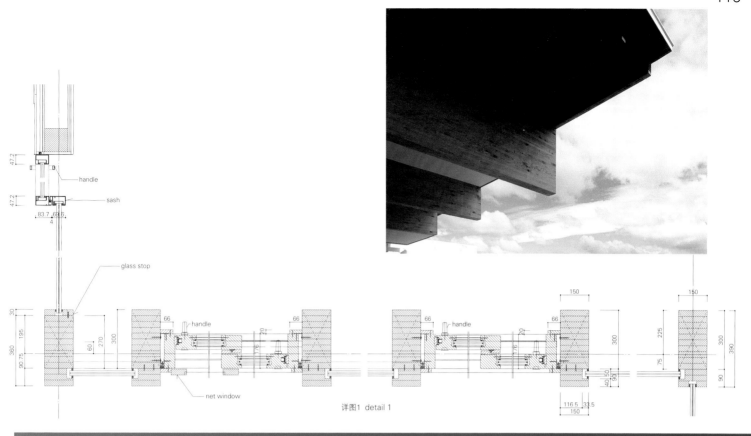

详图1 detail 1

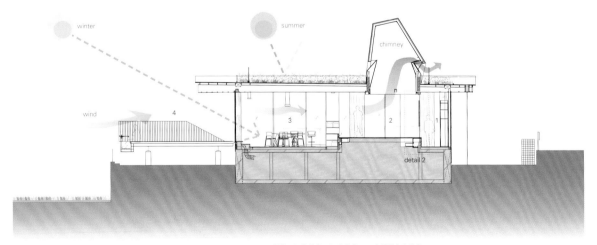

1 走廊 2 会议室 3 办公室 4 木质平台办公室
1. corridor 2. meeting room 3. office 4. wood deck office
A-A' 剖面图 section A-A'

1. large section laminated wood
   made in Aichi Prefecture
   xyladcor paint
2. slate t=6mm
   furring strips t=20mm
   waterproof membrane
   insulating fiber board t=12mm
3. plaster board t=9.5mm
   acrylic emulsion paint
   structural plywood t=12mm
   grass wool t=200
4. plaster board t=12.5mm
   acrylic emulsion paint
   structural plywood t=12mm
5. fiberglass-reinforced plastics waterproofing
   insulating fiber board t=12mm
   structural fiber board t=12mm
   steel frame backing
6. air diffuser

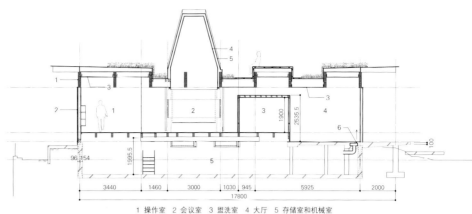

1 操作室　2 会议室　3 盥洗室　4 大厅　5 存储室和机械室
1. operation room　2. meeting room　3. lavatory　4. lobby　5. storage and machine room
B-B' 剖面详图　section detail B-B'

1. main distributing frame t=12mm
2. seamless line
   acrylic emulsion paint
   structural plywood t=12mm
3. projector screen
4. whiteboard
5. plaster board t=12.5mm
   acrylic emulsion paint
   structural plywood t=12mm
6. square pipe steel-50×50×3.2mm
7. steel plate-3.2mm
   steel angle-30×30×3mm@150mm
8. flooring t=15mm
   structural plywood t=24mm
9. base plate
10. steel plate-3.2mm
11. air diffuser
12. supply air chamber
    rigid polyurethane foam t=20mm

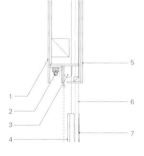

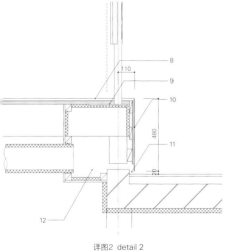

详图2　detail 2

Roduit工作室

Savioz Fabrizzi Architectes

客户在寻求一个50m²左右的绘画和雕刻工作室，该工作室作为现有房屋的一个扩建结构，坐落于新结构几米开外的地面上。

狭小的布局和地方法规对建筑的形式影响很大，这座建筑沿着草地狭窄的长度伸展出来，它不寻常的形状预留了门和窗的位置，为工作室提供了山谷和对面的山峰的宽阔视野。

在外部，工作室在与周围相连的阳台上延伸，形成了与现有房屋视觉上的连接。阳台也包括一处储存区域，带有原地浇筑的混凝土洗手盆和工作台，以便所有类型的工作，例如颜料或其他材料的混合，都能够在外面进行。

该项目也增加了其他房间的开发，例如地下室的木质存储室和小花园储存区域。建筑的入口处沿着陡峭的斜坡设置，使改变尽可能的少，同样的做法也用于屋顶，同样遵循了场地的斜面。

建筑的形状和位置使它好像草地上放置的雕塑品一样，给人以非常深刻的印象。正如图解中看到的，项目已经重新开发不止一次，直到授予建筑许可和施工工作最终开始，在立面上一直结合了混凝土材料和冷杉木使用。

使用的材料突出了岩石般的背景，同时遵守30%木质立面的法规要求。正因为此，使用的混凝土不仅与木材的长度一致，而且预留出箱形凹位，使风化的木质饰面的长度适合斜切到接下来的外部墙体中。这两种材料的流动过渡定义了结构的统一性。窗户也已经嵌装完成，因此建筑的几何结构将更加清晰。

在内部，尤其要注意墙上中性饰面的使用，这也是为什么应用内部保温层的原因，石膏和修补的白灰也加以利用进行覆盖。由于北面的屋顶光，额外的自然光束穿过建筑，也提供了令人印象深刻的附近山顶的景观。储藏架、长椅和水槽在原位浇筑而成，合并了工作室的外部墙体和内部，一起形成了协调一致的整体。

Roduit Studio

The client was seeking a painting and sculpture studio of around fifty square metres as an extension to the existing house and studio situated just a few metres away from the land to be used for the new structure.

The narrowness of the plot and local regulations considerably affected the form of the building which stretches out narrowly along the length of the meadow. Its unusual shape allowed for the positioning of the door and windows to provide great views on the valley and on the mountains opposite.

Outside, the studio extends over a terrace that is related to its surroundings and creates a visual connection with the existing house. The terrace also includes a storage area with a cast-in-place concrete washstand and working surface, so that all kinds of work

Other rooms, such as the wood store room in the basement and the small garden storage area, were added as the developed projects.

The access points to the building were positioned along the steeply-sloping ground so that it was altered as little as possible. The same was applied to the roof, which likewise followed the incline of the plot.

The building's shape and its position create the impression of having placed a sculptural object in the meadow. As it can be seen in the schema the project has been re-developed more than once, and until its building permission was given and construction works could finally begin, it always integrates both materials concrete and fir wood in the facade.

The materials were used to highlight the rocky backdrop, while complying with the regulatory requirement for 30% of the facade to be of wood. For this reason, the concrete used was not only lined with lengths of wood, but also had box-outs enabling the lengths of weathered-finish wood to be fitted flush to the outside wall subsequently. The fluid transition between these two materials defines the unity of the structure. The windows have also been mounted flush so that the building's geometry would be more clearly defined.

Inside, particular attention has been paid to the use of a neutral finish on the walls, which is why they have been insulated on the inside, covered with plaster and fettled white. Additional natural light streams through the building thanks to the northern rooflight, which also provides a view of the impressive nearby mountain top. The storage shelves, benches and sinks were cast in site, uniting the external wall of the studio with its interior to form a harmonious whole.

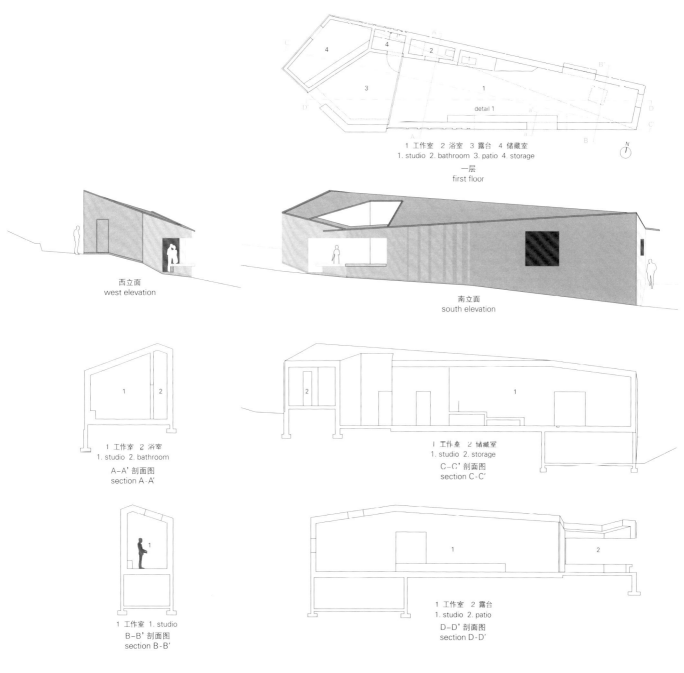

1 工作室 2 浴室 3 露台 4 储藏室
1. studio 2. bathroom 3. patio 4. storage

一层
first floor

西立面
west elevation

南立面
south elevation

1 工作室 2 浴室
1. studio 2. bathroom

A-A' 剖面图
section A-A'

1 工作室 2 储藏室
1. studio 2. storage

C-C' 剖面图
section C-C'

1 工作室 1. studio

B-B' 剖面图
section B-B'

1 工作室 2 露台
1. studio 2. patio

D-D' 剖面图
section D-D'

项目名称：Roduit Studio
地点：Chamoson, Switzerland
建筑师：Savioz Fabrizzi Architectes
项目团队：Benedikt Bertoli-Suelzenfuss, Laurent Savioz
土木工程师：Alpatec Sa
甲方：Josyane and Michel Roduit
功能：artist studio, wood store, garden storage
用地面积：584m²
体积：470m³
总建筑面积：106m²
有效楼层面积：126m²
竣工时间：2011
摄影师：©Thomas Jantscher

roof
- low line eternit roof slate XL vulcanite N6520 40×72cm: 0.5cm
- pinewood roof battens: 3cm
- ventilation/pinewood roof battens: 6cm
- sarnafil TU222 sub-roof membrane stainless steel profile for de-watering: 22×12cm
- pavatex isoroof nature insulation 6cm
- pinewood rafters/glass wool insulation: 22cm
- SIGA majpell 5 steam brake
- battens: 2.7cm
- 2×1.25cm fettled drywall: 2.5cm

ceiling
- reinforced concrete ceiling: 20cm
- jackodur extruded polystyrene foam insulation: 16cm

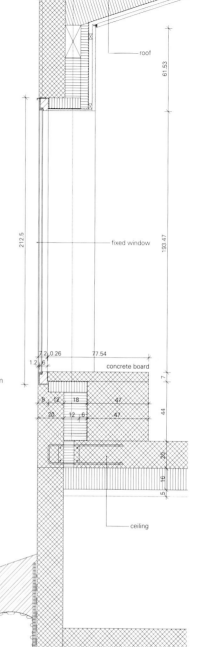

a-a' 剖面图 section a-a'

exterior wall
- 8.5cm vertical pre-weathered white fir: 2.5cm
- horizontal pinewood battens
- ventilation: 2.2cm
- reinforced concrete: 20/26cm
- isover uniroll 035 glass wool insulation: 12+6cm
- isover steam brake
- battens: 2.7cm
- 2×1.25cm fettled drywall: 2.5cm

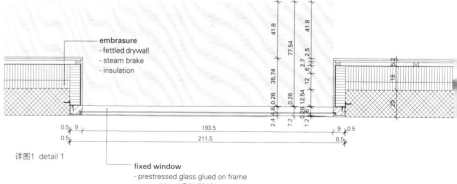

详图1 detail 1

embrasure
- fettled drywall
- steam brake
- insulation

fixed window
- prestressed glass glued on frame
- enamel band RAL 7016
- drip plate matt black anodized aluminium BWB colinal
- interior frame painted white RAL 9010: 6×9cm

木建筑再生
Recovering Wood

一直以来，木材遍及各地，资源丰富，用途广泛，但纵观历史，它作为建筑材料的流行程度却受到不断变化的品味和态度的影响。起初，由于广泛的使用和实验，木材的发展潜能和优势被发掘出来，最近，系统的研究和制造工艺的突破性进步使木材的发展潜能和优势得到进一步印证。我们对环境的担忧进一步促成了人们的发现，即木材是一种可再生资源，提供了可替换的途径，对耗碳的建筑工业形成了挑战。

本章中举例说明的各种项目表明，作为当代建筑材料的木材重新获得了建筑师的青睐，揭示了建筑师对它重新燃起兴趣的原因。他们触及大量建筑体系之内反复出现的主题，就像我们将自己的立场与"自然"相对比，告诉我们，木材的使用可以缓和自然环境和人造环境之间的巨大差异。这些项目强调我们与自然的长期关系，强调木材的存在对我们身心健康的影响。它们揭示了科技成就将传统木工技艺转化成动态实践的过程，帮助我们了解文化和建筑如何互相影响。木材重获青睐，但同样重要的是，它们是作为我们城市中心的主要建筑材料而回归的。

Although wood has always been around as a resourceful material for all sorts of creations and purposes, its popularity as a building material has been under the influence of changing tastes and demeanours throughout history. The possibilities and merits of wood were at the start exposed through extensive use and experimentation, and more recently by systematic research and the advance of ground-breaking manufacturing processes. Concerns about our environment also precipitated the discovery that wood is a renewable resource, and offers alternative routes to challenge our generally carbon-hungry construction industry.

The projects illustrated within this chapter show the recovery of wood as a contemporary construction material and reveal a wide array of motivations for this renewed interest. They touch upon a number of reoccurring themes within architecture like our position versus "nature", and show us that the application of wood can soften the dichotomies between our natural and built environment. The projects emphasize our longstanding relation with nature and the impact the presence of wood has on our health, both psychologically and physiologically. They demonstrate how technological discoveries turned our traditional carpentry into a dynamic practice and helped us understand how culture and construction condition one another. Wood is back in fashion and last but not least is making its return as a key construction material at the very centre of our cities.

Tamedia办公大楼_Tamedia Office Building/Shigeru Ban Architects
主教爱德华国王礼拜堂_Bishop Edward King Chapel/Níall McLaughlin Architects
杰克逊霍尔机场_Jackson Hole Airport/Gensler
皮塞克市林业局_Pisek City Forest Administration/HAMR
园艺展览中的展馆_Pavilion for Horticultural Show/Dethier Architecture
Reussdelta瞭望塔_Reussdelta Observation Tower/Gion A. Caminada

木建筑再生_Recovering Wood/Tom Van Malderen

自人类开始建造遮风挡雨的住所以来，木材一直是一种重要的建筑材料。从人们认识到他们可以使用不同材质的那一刻起，他们就一直使用木材，琢磨它，把它转化成用途最广泛的建筑材料之一。起初，木材种类丰富，不需要精细的工具对它进行加工，即使加工也只是为了使它更受欢迎。在接下来的几个世纪里，人们不断塑造世界，不断建造大楼、船舶、车辆、机器和风力发动机等等；同时，世界人口不断增长，需要增加更多的农田来消除饥饿，结果使得现存的森林受到威胁，变得异常珍贵。

古往今来，人们喜爱木材的程度一直受到时尚、品味和不断变化的评判标准的影响。举例来说，在18世纪，建筑材料的承受能力在很大程度上与建筑有关，似乎要坚定稳固如埃及的金字塔，很明显，人们喜欢使用石材。这种历史和文化的成见一直持续至今。虽然木材有时因看上去缺乏"坚定稳固"感而不被看好，但在现有的环境中，木材表现出侵入性较小的特质，给人感觉更温和、更柔软。这些情感因素在本章强调的各种项目中起到了绝对作用。

对木材属性和品质的理解最初来自经验，后来来自系统的研究和先进的检测。随着时间的推移，木材已经从一种简单的、现成的天然材料演变成一种现代工业和工程材料。我们不仅发现了木材在热学、声学、电学、机械和审美方面的品质，而且全球变暖问题和环境问题也让我们意识到木材的碳汇效益。不像一般耗碳的建筑业中使用的大部分材料，木材作为一种从可再生资源中获得的建筑材料，生产时能量消耗相对较少，而且几乎100%可回收利用、可生物降解。

在过去的十多年里，研发了许多可替换的建筑材料，木材面临着许多竞争，但是，我们可以看到建筑师重新燃起了对木材的兴趣：对已有传统的兴趣和源于木材新创意和后续发展潜能的兴趣。本章中的项目清楚地证明了木材重归时尚，受到青睐。这种回归的考虑涉及多个层面，受到技术和意识形态的积极影响，以长期存在的建筑传统、建筑理论和建筑主题为基础形成。

其中的一个主题是将人的立场与自然相对比，通常情况下，即我们对自然的感知和定义，以及随之而来的巨大差异。"真正的自然"依然存在吗？还是给我们留下了有深远意义的、绿色的和谐世界的一种怀旧的、可能理想化的、风景如画的形象？是一个已经失去却又渴望得到的世界吗？我们应该把自己看作是自然的一个部分呢，还是自然之外的呢？我们往往认为自然代表着美好与善良，但既然自然会出现井然有序和暴力混乱的情形，那么"自然之美"在概念上难道没有预示危险吗？是否可以通过生物工程，找寻"智能材料"，来形成一种"工业"和"自然"

Wood has been an important construction material since humans began building shelters. From the first moment people recognized they could make use of materials, wood has been used and improved and turned into one of the most versatile and useful construction materials for many years. At first, wood was available in abundance and the fact that no elaborate tools were needed to work it only added to its popularity. Centuries of shaping the world followed, with the continuous fabrication of buildings, ships, carts, machines, windmills, etc., all the while an ever growing world population increased the demand for more farmland to satisfy its hunger. So much so that today's remaining forests have turned into precious and threatened environments.

The popularity of wood through the ages has also been influenced by fashion, taste and changing judgements. To illustrate with an example, in the 18th century, the structural competence of building materials was very much linked to architecture that appeared to be as eternally stable as the Egyptian pyramids, and clearly favoured the use of stone. Some of these historic and cultural preconceptions keep lingering on till today. Whilst wood is sometimes being looked down upon as lacking that "monumental appearance of stability", it equally acquired the attribute of being less invasive or aggressive to an existing context; of being gentler and softer. These sentiments are definitely at play in a number of the projects highlighted within this chapter.

The properties and qualities of wood were initially understood by experience, and more recently by systematic research and advanced examination. Over time, wood has evolved from a simple, readily available natural material to a modern industrial and engineering material. We have not only discovered the thermal, acoustic, electrical, mechanical and aesthetic qualities of wood, but global warming issues and environmental concerns also made us realize that wood is a carbon-sink. Unlike most of the materials used in our generally carbon-hungry construction industry, wood is a construction material harvested from renewable resources, manufactured with relatively low energy input, and almost 100% recyclable and biodegradable.

Although wood received a lot of competition from the development of many alternative construction materials over the last decennia, we can see a remarkable resurgence of interest in wood: an interest directed both towards previously established traditions and towards new inventions and the subsequent possibilities for wood. The projects shown here are a clear demonstration that wood is back in fashion and on the recovery. This renewed consideration is multi-layered, stimulated by technology and ideology, and building upon longstanding traditions, theories and themes within architecture.

One such theme is man's position versus nature, our perception and definition of nature in general and the many dichotomies that come along with it. Does "real nature" still exist or are we left with a nostalgic and possibly idealised form of picturesque image of a profoundly green, harmonious world? A world that is both lost and desired again? Should we see ourselves as part of

园艺展览中的展馆,科布伦茨市,德国,2011年
Pavilion for Horticultural Show, Koblenz, Germany, 2011

的综合体?我们的责任与自然相比会怎样呢?与越来越让人绝望的环境报告而产生的内疚和焦虑的感觉相比会怎样呢?我们并没有强调人与自然之间的巨大分界线,或者反对将自然界与本章中的部分人造项目相对比,相反,我们将木材作为主要的建筑材料,从而使自然环境与人造环境之间的过渡更加模糊。Dethier建筑事务所设计的园艺展览中的展馆是一座木质的三角形观景台,是观赏自然风景、科布伦茨城市风景和观景台自身的最佳地点。建筑师将现代高科技设计与周围的公园景观和山谷中的城市有机结合,他们想通过对自然的感知来给城市灌注一种新的象征,从而创建一种视觉信标,同时使其和谐地融入山谷的风景。

同样,木材作为一种活跃的元素,使Gensler建筑事务所设计的杰克逊霍尔机场和HAMR设计的皮赛克市林业局与周围的自然环境几乎融为一体。正如前面所解释的,他们把木材当作一种侵入性较小的材料,在现有的环境中使用它,试图将人类文明的痕迹降到最低,甚至暗示这只是临时性的。为了尽可能不要分散注意力,两个项目的建筑师在他们的结构中安装了大型平面玻璃,将视线从他们自己的建筑上转移开,以突显人类、自然和建筑物之间的亲密关系。

让人身体健康的环境取决于多个方面,其中包括随时呼吸到充足的新鲜空气,沐浴丰富的自然光线,保持舒适的温度和湿度,以及尽可能少接触的污染物。但是,随着时间的推移,可持续建设的定义也随之丰富,因此,我们对人类健康的理解也扩大了,即人类健康不仅包括我们的身体健康,也包括我们的心理健康。虽然不是什么伟大的奇闻轶事,但马丁·海德格尔和路德维希·维特根斯坦是在木屋里完成了他们20世纪的部分伟大哲学著作。他们都喜欢在本土的木屋里度过时光,因为那里可以让他们保持清醒的头脑,能够激发他们的想象力。

很长一段时间以来,我们就知道人们与自然的关系亲密,在自然环境中可以让我们感到更加平静。此外,最近的研究已经证实,自然的存在可以减轻我们的神经系统面对压力的反应。在20世纪80年代,爱德华·O.·威尔逊介绍并推广了"亲生命性"这一术语:亲生命性假说表明,人类和其他生物系统有一种本能的关系。有明显迹象表明,树木的存在触发了与自然的这种联系,从而产生了积极的情感。事实上,许多人在情感方面会对树木产生回应,认为它很"温暖",这似乎找到了科学依据,并暗示了树木与人类心理及生理健康之间的有益联系。

这种联系在Niall McLaughlin建筑师事务所设计的主教爱德华国王礼拜堂里得到完美体现。内部的木质结构形成了纵横交错的拱梁,让

nature, or above it? And whilst we tend to refer to nature as a sign of the good and virtuous, isn't the "beauty of nature" fraught with conceptual danger, since nature is both emergent order and violent chaos? Can there be a synthesis of the "industrial" and the "natural" through bio-engineering and looking for "smart materials"? And what about our responsibility versus nature, the feeling of guilt and anxiety as a result of increasingly dooming environmental reports? Rather than underlining a great divide between man and nature, or opposing the natural versus the artificial in some of our chapter's projects, instead we blur the transition between natural and built environment with the application of wood as the predominant building material. Pavilion for Horticultural Show by Dethier Architecture is a wooden triangle that offers views of the landscape, the city of Koblenz and the belvedere structure itself. The architects unite a contemporary high-tech design with the surrounding park landscape and the city in the valley. They want to reconcile a new symbol for the city with the perception of nature, create a visual beacon and simultaneously harmoniously blend.it into the valley's landscape.
Similarly, the Jackson Hole Airport by Gensler and the Pisek City Forest Administration by HAMR apply wood as an active element to diminish the threshold with the natural environments they are surrounded by. As previously explained, they use wood as a less invasive material towards the existing context. They try to keep the traces of human civilization to a minimum, maybe even hint at being of a temporary nature. In order to distract even less, both architects opened up their structures with large planes of glazing to direct the view away from their own constructions in order to stimulate an intimate connection between human, nature and building.
Many aspects contribute to a physically healthy environment, including sufficient fresh air changes, abundant natural light, keeping comfortable levels of temperature and humidity and limited exposure to pollutants. But just as the definition of sustainable building has expanded with time, so has our understanding of human health expanded to include not only our physical state, but our psychological wellbeing as well. It may be a small anecdote, but both Martin Heidegger and Ludwig Wittgenstein prepared some of their great twentieth century philosophical works in wooden huts. They both loved to spend time in a vernacular timber cabin where they could clear the mind and stimulate their imagination.
We have known for a long time that people have an affinity for nature, and that being in a natural environment can make us feel more tranquil. Moreover, recent studies have confirmed that the presence of nature reduces stress reactions in our nervous system. In the eighties Edward O. Wilson introduced and popularized the term "biophelia": the Biophilia Hypothesis suggests that there is an instinctive bond between human beings and other living systems. There are clear indications that the very presence of wood triggers this connection with nature and generates positive feelings accordingly. The fact that many people emotionally respond to wood and refer to it as something "warm" seems to find scientific reasons, and hints at a beneficial link between wood and

Tamedia办公大楼，苏黎世，瑞士，2013年
Tamedia Office Building, Zurich, Switzerland, 2013

我们想起了哥特式圆柱与其灵感来源：森林里纵横交错的高大树枝。但同样的，礼拜堂周围的自然环境充分体现了礼拜堂内部的特点，Niall McLaughlin因此解释道："如果你起得非常早，在日出的时候，太阳从地平线上升起，会将不断移动的树枝、树叶、窗棂和结构的影子投射到天花板上，纵横交错，很像在树林里仰望树木。"他们建造了一种精美的木质结构，延伸至树梢，表明了乐观向上的精神，最后走向了光明。

木材的使用也为建筑和当地传统之间构建了一座桥梁。Reussdelta瞭望塔的建筑师Gion A. Caminada在提出了这样一个问题——这些古老的建筑体系可能带给我们什么启示时，说道："这些建筑的实质体现在哪里？它们怎么就能因为新的用途而得以转变以满足现代的需求呢？"他在瑞士Vrin市为鸟类学家设计建造的瞭望塔就是非常具有代表性的建筑之一，以优化Vrin市的作用，推动当地经济，同时还要探究当地的传统和特殊之处。将传统和现代方式有机地联系起来是建筑发挥其社会作用的重要组成部分。了解文化观念如何影响建筑设计，建筑设计如何影响建筑观念在更加广阔的文化想象中的发挥是十分重要的。

现今，世界不再那么清晰地划分为机械制作的倡导者、物品数字化处理的推广者和手工艺传统的热衷者。我们的时代更能体现精微玄妙之处，我们的价值观复杂多变。Shigeru Ban设计的Tamedia办公大楼体现出与传统木工技艺相脱离的特点。其主要结构体系完全由木材组成，完美地体现了日本传统的木工技艺，根本不用接合的器具和胶剂。尽管如此，该建筑是创新之作，四层互锁的木质节点网的周围都有玻璃幕墙环绕着。建筑工作的规模需要对材料进行细致的研究，需要新的建造流程帮助，以获得所需的性能标准。为避免结构上出现任何额外的钢筋，所有预制组件都需要电脑数控铣削技术的精准度。

无论是木材的温暖之美、审美之美、生态宜居还是技术潜力，这种材料必将在可持续的建设和建筑业中发挥重要的作用。作为许多当代项目之一，Shigeru Ban的项目向我们展示了木材在当代城市中心作为主要建筑材料开始回归了。木材已成为一种动态元素，正在复苏：引进了层压、铣削、合并、铸造和组合等先进技术，以至于基于木材的解决方案正在改变建筑的前景。作为一种可再生的建筑材料，其广泛的用途使复杂的空间探索成为可能，现在似乎木材做材料的时刻到了。

human health, both psychologically and physiologically.
This connection is very present in Bishop Edward King Chapel designed by Níall McLaughlin Architects. The internal wooden structure develops into overlapping arched beams, reminding us of Gothic columns and the manner in which they were inspired by the overlapping of tall branches in a forest. But equally is the natural environment surrounding the chapel captured inside, as Niall McLaughlin explains: *"If you get up very early, at sunrise, the horizontal sun casts a maze of moving shadows of branches, leaves, window mullions and structure onto the ceiling. It is like looking up into trees in a wood."* They constructed a delicate timber structure that rises into the treetops to suggest an uplifting buoyancy, rising towards the light.
Employing wood also allows architecture to engage in a conversation with local traditions. Gion A. Caminada, the architect of the Reussdelta Observation Tower raises the question what those old architectural systems may give us: "Where is the substance of those constructions? How can they be transformed for new kinds of use to meet the requirements of the present time?" His observation tower for ornithologists in the city of Vrin in Switzerland is one of many constructions in order to optimize Vrin's functioning and advance the local economy whilst exploring the traditions and particularity of the place. Bridging tradition and modern ways is a crucial component for architecture to play a social role. It is essential to understand how cultural ideas condition construction and how construction, in turn, conditions the play of architectural ideas in a larger cultural imagination.

Today, the world is not so clearly divided into protagonists of machine-made goods, promoters of digitally processed artefacts, and devotees of the handcrafts tradition. Our era is more nuanced, and our value systems are complex and varied. The Tamedia Office Building by Shigeru Ban finds its point of departure with traditional carpentry. Its main structural system is made entirely of timber and is, in the tradition of the Japanese mastery of carpentry, entirely devoid of joint hardware and glue. Nonetheless, the building is an innovative, contemporary four-story interlocking wooden network of nodes enveloped by a glass skin. The scale of the work required a close study of the materials and the help of new fabrication processes to obtain the performance criteria needed. In order to avoid any additional steel reinforcements in the structure, all prefabricated components required the precision of CNC milling technology.
Whether it is wood's warm, aesthetic beauty, its ecological friendliness or technological potential, this material is bound to play a prominent role in a sustainable building and construction industry. Shigeru Ban's project is one of many contemporary projects showing us that wood is also making its return as a primary building material in the very centre of the contemporary city. Wood has become a dynamic element on the recovery: technological discoveries introduced laminating, milling, combining, casting and weaving, so much so that timber-based solutions are changing the outlook of construction. As a renewable building material whose versatility allows for intricate spatial exploration, wood again seems to be one material of the moment. Tom Van Malderen

Tamedia办公大楼
Shigeru Ban Architects

为瑞士传媒公司Tamedia总部设计的新办公楼最近已经竣工。该项目占地1000m²，位于苏黎世市中心一个更大的城区内，主要的建筑群都位于其中。建筑物面向街区东部，近50m长的直线形立面面向斯尔运河，而大楼穿过立面延伸。

新建筑与即将拆毁的原有建筑的轮廓相呼应，但与旁边建筑物的外立面连接到一起，并且利用了最大允许高度，以使建筑街区一侧的可用办公区域得到最优化利用。

建筑的主通道位于Werdstrasse街和Stauffacherquai街北角，这实际上是整座综合建筑设施的主入口。

建筑物共有七层楼，其中地下设有两层，总净面积为8602m²，设计师又额外扩建了1518m²，以响应搭建在Stauffacherquai街8号邻近的建筑屋顶的二层扩建工程。

从建筑学的角度来看，这项工程的主要特点之一是完全由木材构成主要结构体系的构想。同时，源自技术和环境立场上的创新特性，赋予了建筑无论从内部空间还是从周围城市来看都是独一无二的外观。

为了强调和表达这种观点，建筑表皮完全采用玻璃制成。并且，为响应瑞士最新出台的关于能耗方面非常严格的规定，建筑师在实现低能量传输水平方面给予了特别关注。

面向城市，这座建筑还有一个"中间媒介"空间，贯穿东立面的整体高度，它的另一个作用是在总体能量消耗策略中充当"绝热板"，这也成为一处带来独特空间体验的休息区，以及不同办公楼层之间的垂直连接。

这些"阳台"可用作非正式会议区或者休息区，其特殊性在于有一个立面由一个玻璃材质的可伸缩窗口系统组成，能够将这些空间转变为室外露台，以促进内部建筑物及其周围风景之间的亲密关系。

木质的主结构体系在很大程度上是这项工程的最有意义的创新。从技术和环境的角度出发，提出这个木质结构的构想正是对这种类型的办公楼的一个独特回应，而事实上结构元素的完全可见性也提供了一种具有独一无二特性及高品质空间的工作氛围。

除了选择木材作为主结构材料对可持续发展做出的显著贡献（只选用可再生建筑材料，并且将建设过程中的二氧化碳排放量降到最低）外，全球机械系统也被设计应用以满足能源问题的最高标准。中空部位作为公共区域的一部分起到"隔热层"的作用，通过抽出办公区的空气来调节温度。

Tamedia Office Building

The project for the headquarters of the Swiss media company Tamedia is situated in the heart of the city of Zurich in a 1,000m² site within a larger urban block where the main buildings of the group are currently located. The site is positioned towards the east part

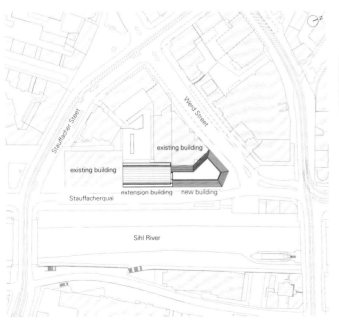

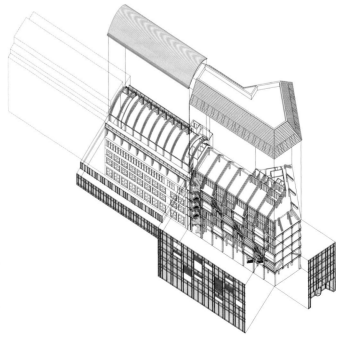

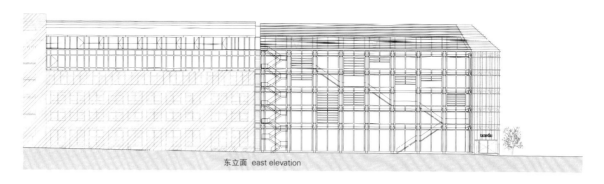
东立面 east elevation

of the block and has the particularity of developing through almost 50m of linear facade facing the Sihl Water Canal.

The implantation of the new building basically responds to the footprint of the existing building to be demolished but this time creates continuity with the facades of the buildings beside as well as takes advantage of the maximum allowed height in order to optimize the exploitable office area in this side of the building block.

The main access of the building is situated in the north angle of streets of Werdstrasse and Stauffacherquai and will actually become the principal entry of the whole building complex.

The building develops within 7 stories over ground floor with two basement levels for a global net area of 8,602m² to which the architects can add 1,518 additional square meters that correspond to the two-floor extension project located on the roof of the neighbor building at number 8 of Stauffacherquai Street.

From an architectural point of view, one of the main features of the project is indeed the proposition of a main structural system

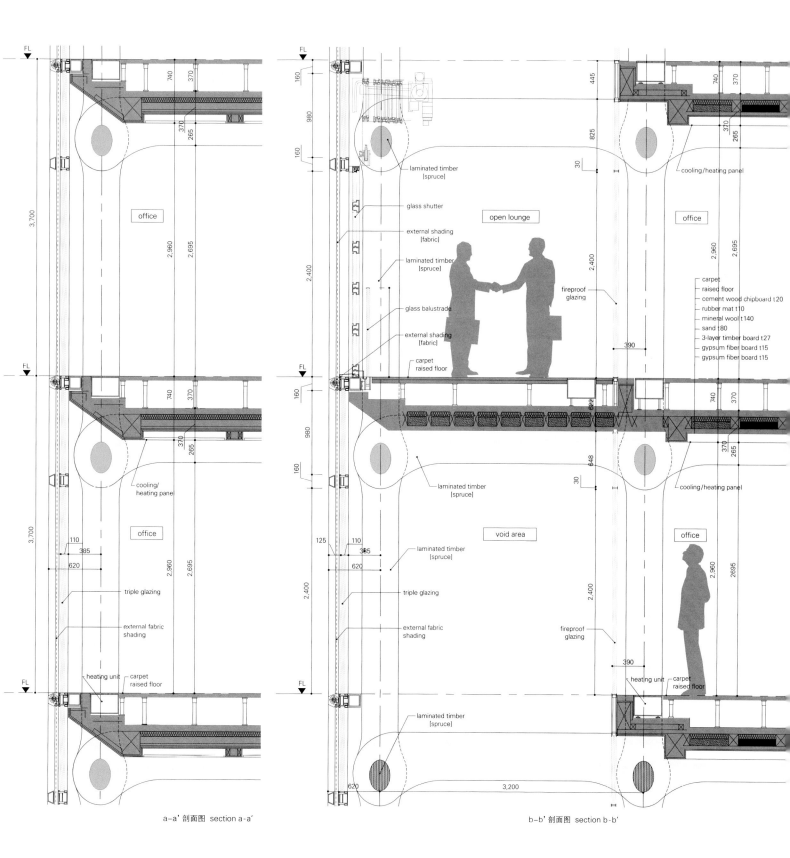

a-a' 剖面图 section a-a'   b-b' 剖面图 section b-b'

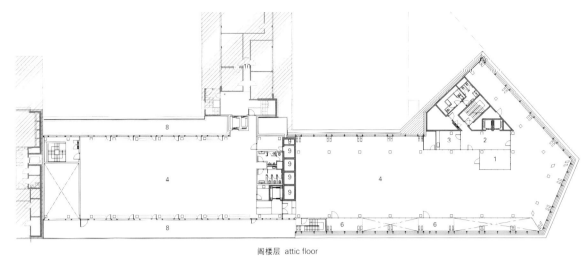
阁楼层 attic floor

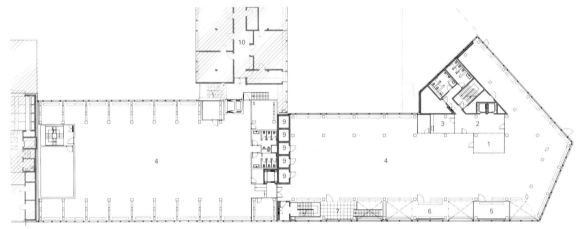
五层 fifth floor

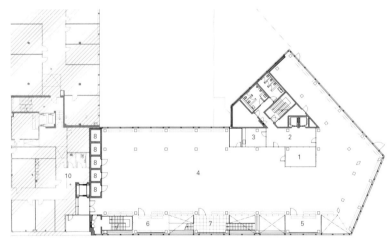
典型楼层平面（二~四层） typical floor plan (second floor ~ fourth floor)

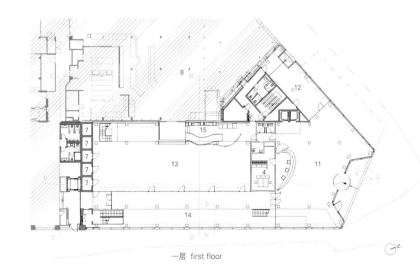
一层 first floor

1 会议室
2 电梯厅
3 厨房
4 办公室
5 开放的休息室
6 封闭的休息室
7 楼梯平台
8 露台
9 管道空间
10 现存建筑
11 入口大厅
12 租用空间
13 多功能空间
14 休息室
15 厨房角落

1. meeting room
2. elevator hall
3. kitchen
4. office
5. open lounge
6. closed lounge
7. landing
8. terrace
9. duct space
10. existing building
11. entrance lobby
12. tenant space
13. multipurpose space
14. lounge
15. kitchen corner

entirely made of timber. And it has an innovative character from a technical and environmental standpoint, giving the building a unique appearance from the inside space as well as from the city around. In order to reinforce and express this idea, the building's skin was entirely glazed and special attention was given to achieve a low energy transmission level that responds to the latest and very strict Swiss regulations in terms of energy consumption. Facing the city, the building also features an "intermediate" space throughout the whole height of the east facade that plays its role as "thermal screen" within the general energy consumption strategy, and also becomes a unique spatial experience with lounge areas and vertical links between the different office stories. These "balconies" can be used as informal meeting and rest areas that will also have the particularity of having a facade composed of a glass retractable window system that allows to "transform" these spaces into open air terraces that reinforce the privileged relationship between the interior building and its surrounding landscape. The timber main structural system is to great extent the most significant innovation of the project. From a technical and environmental point of view the proposed timber structure is a unique response to this type of office building and the fact that the structural elements are entirely visible also gives a very special character and high quality spatiality to the working atmosphere.

Besides the clear contribution to sustainability on the choice of timber as the main structural material (only renewable construction material and the lowest CO_2 producer in construction process) the global mechanical system has been designed to meet the highest standard in energy issues. The intermediate space which functions as "thermal barrier" is part of the public space that will be heated and cooled with the extraction air from the office area.

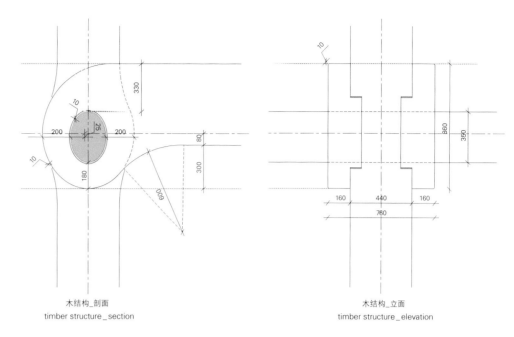

木结构_剖面
timber structure _ section

木结构_立面
timber structure _ elevation

项目名称：Tamedia Office Building
地点：Zurich, Switzerland
首席建筑师：Shigeru Ban
合作建筑师：Jean De Gastines
项目建筑师：Kazuhiro Asami, Gerardo Perez, Takayuki Ishikawa, Masashi Maruyama
本地建筑师：Itten+Brechbuhl AG
结构工程师：Creation Holz GmbH
MEP工程师：3-Plan Haustechnik AG
总承包商：HRS Real Estate AG
甲方：Tamedia AG
用地面积：1,000m² (总用地面积：8,000m²)
总建筑面积：1,000m²
有效楼层面积：10,120m²
结构：timber structure, reinforced concrete
设计时间：2008.4~2010.12 / 竣工时间：2013.3
摄影师：©Didier Boy de la Tour (courtesy of the architect)

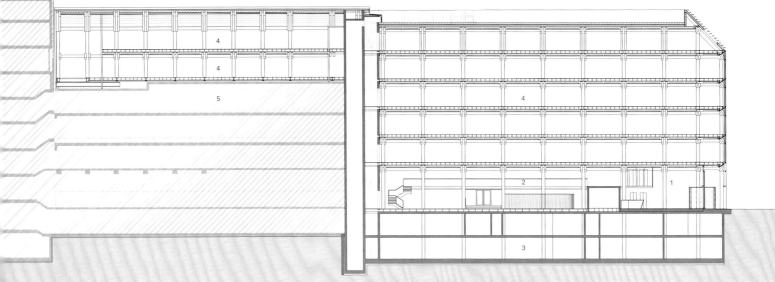

1 入口大厅 2 储藏室/技术室 3 多功能空间 4 办公室 5 现存建筑
1. entrance lobby 2. storage / technical room 3. multipurpose space 4. office 5. existing building
A-A' 剖面图 section A-A'

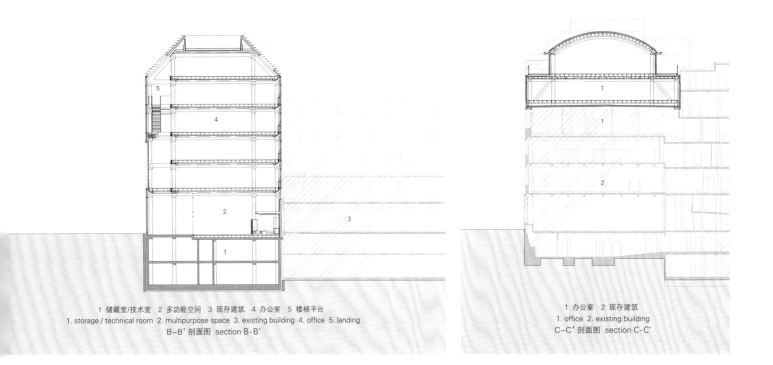

1 储藏室/技术室 2 多功能空间 3 现存建筑 4 办公室 5 楼梯平台
1. storage / technical room 2. multipurpose space 3. existing building 4. office 5. landing
B-B' 剖面图 section B-B'

1 办公室 2 现存建筑
1. office 2. existing building
C-C' 剖面图 section C-C'

甲方的项目书称想要为里彭神学院寻找一个新的礼拜堂，为住在牛津郡校园两个相互关联的群体服务。一个群体为学生，另一个群体为小型宗教组织的修女们——博格布洛克的姐妹。新教堂将取代现有的教堂，那是19世纪晚期由乔治·埃德蒙·斯特里特设计的，如今因其太小已不能满足实际需要。项目书要求新教堂能够满足大学内这两个团体做礼拜的需要，并且也适合公共聚会和个人礼拜。此外，项目书还设想了一个专门供博格布洛克的修女们背诵的独立空间，一间宽敞的圣器安置室和一间必要的附属房间。在这些大致要求之余，项目书也表明了甲方希望教堂最重要的是能够作为个人与圣灵相遇的地方，可以让个体创造性地思考空间和礼拜仪式之间的关系。

山眉上有一棵巨大的山毛榉，在背向山毛榉以及教学楼后面的地方，绿树在高地环绕，俯瞰着山谷，山谷一直向Garsington延伸。学校现有的建筑具有重要的历史意义。这个地方在南牛津郡地方计划中被指定为城市绿化带，它穿过峡谷一直向西延伸很远，人们从很远的地方就能看见它。学校附近大树参天，并且在东部边界实行树木保护令，新礼拜堂的设计既要融入全景的特征又要保留校园及周围树木的背景。对于这些计划中相互交叠的敏感问题，需要与南牛津郡区委员会、英国古迹署以及当地居民进行广泛协商。

主教爱德华国王礼拜堂
Níall McLaughlin Architects

谢默斯·希尼的诗歌《阳光》(8) 中的"教堂的中殿"一词正是这一项目的出发点。这个词的意思是教堂的中心，同时又与"船"这一词同源，此外它还可以指转动的车轮中静止的中心部位。

这些词语形成两种建筑意象。第一个是在地面上的中空空间，即静止的中心，作为群体的集会场所。第二个是漂浮于树冠之上，似船的木质结构，是灯光和音响的交集场地。教堂四周的回廊呈现出几何图形固有的运动轨迹。绕堂而行，可见中心宽敞明亮。窥探发光的空地，这种感觉可以追溯到最早期的教堂。

从外部看，教堂是一座简单的由石头围住的围场。建筑师使用了克利普舍姆石，这种石头在纹理和色泽上都与学校现有的石灰岩材质相呼应。外墙是隔音腔结构，其内层是弯曲的钢筋砌块片，而外层是细琢石片。教堂墙壁由大量的天然石质翅片砌成。翅片位于高性能双层玻璃的前臂，安装在隐蔽的金属框架里。

主教堂及附属楼的屋顶都采用了具有保温性能的板材结构。教堂屋顶的排雨管隐藏在外墙的隔音腔内。凡在与天窗齐平的位置，排雨管都装在带有电镀黄铜饰面的铝管里，再嵌入石质翅片中。屋顶和内部框架各自独立，同时独立于外墙，这一点从屋顶与外墙的最小的节点便可看出。

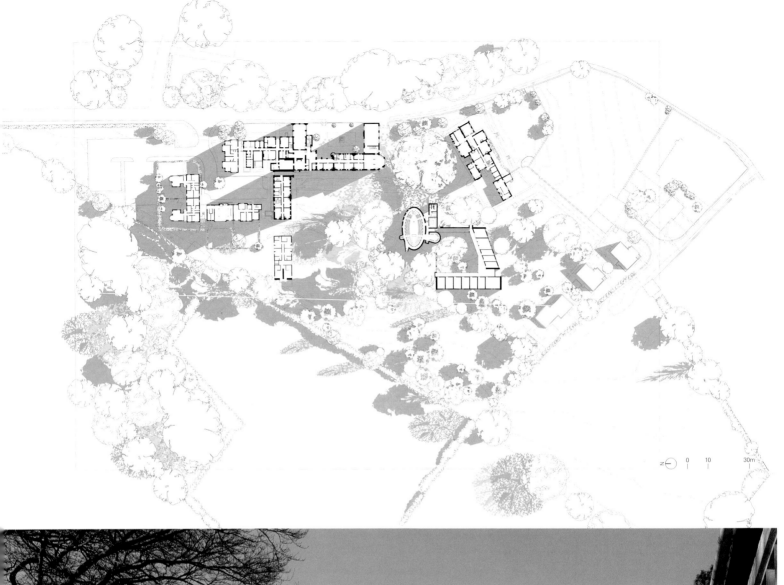

教堂内部的木质结构采用的是预制胶合木（用钢结构固定），以及完全隐藏的钢底板连接构件。该胶合木是由经过染色处理的云杉层压木制成，带有明亮的白涂料饰面。屋顶和柱子的结构体现了椭圆本身的几何结构：在中心与边缘之间直线往返，两头恰是建筑焦点所在，而这也反映出学校的另外两个焦点，即讲台和圣坛。

当人们在教堂周围行走时，柱群和形式简单的椭圆形墙体之间富有韵律的互动便呈现出来。

Bishop Edward King Chapel

The client brief sought a new chapel for Ripon Theological College, to serve the two interconnected groups of residents on the campus in Oxfordshire, the college community and the nuns of a small religious order, the Sisters of Begbroke. The chapel replaces the existing one, designed by George Edmund Street in the late nineteenth century, which had proved to be too small for the current needs of the college. The brief asked for a chapel that would accommodate the range of worshipping needs of the two communities in a collegiate seating arrangement, and would be suitable for both communal gatherings and personal prayer. In addition the brief envisioned a separate space for the Sisters to recite, a spacious sacristy, and the necessary ancillary accommodation. Over and above these outline requirements, the brief set out the clients' aspirations for the chapel, foremost as "a place of personal encounter with the numinous" that would enable the occupants to think creatively about the relationship between space and liturgy.

On the site is an enormous beech tree on the brow of the hill. Facing away from the beech and the college buildings behind, there is a ring of mature trees on high ground overlooking the valley that stretches away towards Garsington. The college's existing buildings are of considerable historical importance. The site is designated within the Green Belt in the South Oxfordshire Local Plan and is also visible from a considerable distance across the valley to the west. The immediate vicinity of the site is populated with mature trees and has a Tree Preservation Order applied to a group at the eastern boundary. The design needed to integrate with the character of the panorama and preserve the setting of

1 入口大厅
2 主礼拜堂
3 回廊
4 用于圣餐的壁龛
5 私人祈祷室
6 博格布洛克姐妹的祈祷室
7 圣器收藏室
8 卫生间
9 储藏室
10 钟楼
11 次入口
12 群体祈祷室

1. entrance lobby
2. main chapel
3. ambulatory
4. blessed sacrament niche
5. private prayer space
6. Sister's prayer room
7. sacristy
8. toilets
9. storage
10. bell tower
11. second entrance
12. prayer board

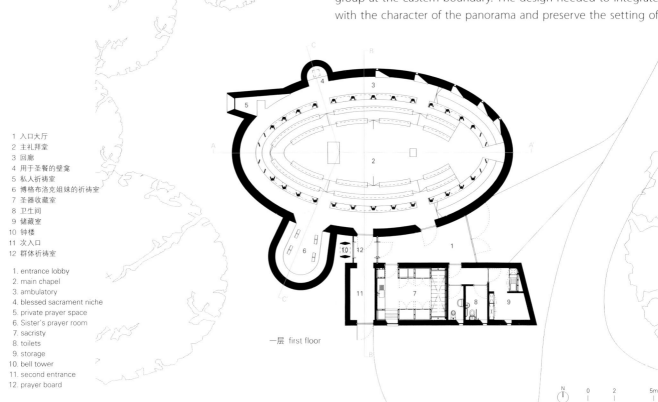

一层 first floor

the college campus and the surrounding trees. The mediation of these interlocked planning sensitivities required extensive consultation with South Oxfordshire District Council, English Heritage and local residents.

The starting point for this project was the word "nave" at the centre of Seamus Heaney's poem *Lightenings viii*. The word describes the central space of a church, but shares the same origin as "navis", a ship, and can also mean the still centre of a turning wheel. From these words, two architectural images emerged. The first is the hollow in the ground as the meeting place of the community, the still centre. The second is the delicate ship-like timber structure that floats above in the tree canopy, the gathering place for light and sound. The movement inherent in the geometry is expressed in the chapel through the perimetric ambulatory. It is possible to walk around the chapel, looking into the brighter space in the centre. The sense of looking into an illuminated clearing goes back to the earliest churches.

The chapel, seen from the outside, is a single stone enclosure. The architects have used Clipsham Stone which is sympathetic, both in terms of texture and colouration, to the limestone of the existing college. The external walls are of insulated cavity construction, comprising of curved reinforced blockwork for internal leaf and dressed stone for outer leaf. The chapel wall is surmounted by a halo of natural stone fins. The fins sit in front of high-performance double glazed units, mounted in concealed metal frames.

The roof of the main chapel and the ancillary block are both of warm deck construction. The chapel roof's drains to conceal rainwater pipes run through the cavity of the external wall. Where exposed at the clerestory level, the rainwater pipes are clad in aluminium sleeves with a bronze anodised finish and recessed into the stone fins. The roof and the internal frame are self-supporting and act independently from the external walls. A minimal junction between the roof and the walls expresses this.

The internal timber structure is constructed of prefabricated glulam sections with steel fixings and fully concealed steel base plate connections. The glulam sections are made up of visual grade spruce laminations treated with a stain system, which gives a light white-washed finish. The structure of roof and columns express the geometrical construction of the ellipse itself: a ferrying between centre and edge with straight lines that reveals the two stable foci at either end, reflected in the collegiate's layout below in the twin focus points of altar and lectern. As you move around the chapel there is an unfolding rhythm interplay between the thicket of columns and the simple elliptical walls beyond.

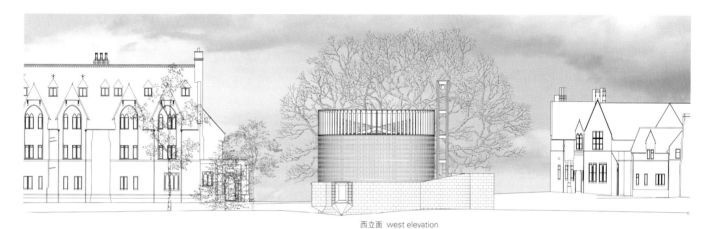

西立面 west elevation

南立面 south elevation

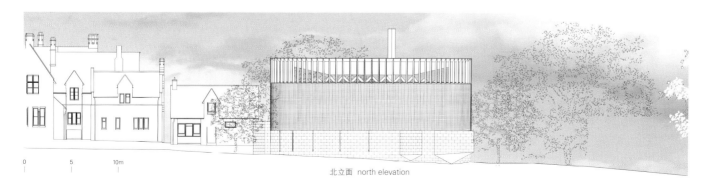

北立面 north elevation

1 木框架
1. timber frame

2 框架支撑着木质屋顶
2. frame supports timber roof

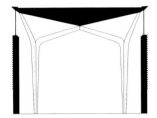

3 墙体包围着框架
3. walls encircle frame

椭圆形窗户平面详图
elliptical window plan detail

项目名称：Bishop Edward King Chapel
地点：Oxford, England
建筑师：Níall McLaughlin Architects
承包商：Beard Construction
结构工程师：Price and Myers
M&E工程师：Synergy Consulting Engineers
音效工程师：Paul Gillieron Acoustic Design
工料测量师：Ridge and Partners LLP
石材顾问：Harrison Goldman
接入顾问：Jane Toplis Associates
规划顾问：Nathaniel Lichfield and Partners
CDM协调：HCD Management Limited
建筑控制审查员：HCD Building Control
施工顾问：Richard Bayfield
甲方：Ripon College and Community of St John the Baptist
有效楼层面积：280m²
竞赛时间：2009.7 / 规划时间：2010.6
施工时间：2011.7 / 竣工时间：2013.2
摄影师：
courtesy of the architect-p.140, p.146, p.148, p.149
©Nick Kane-p.138~139, p.142, p.144~145

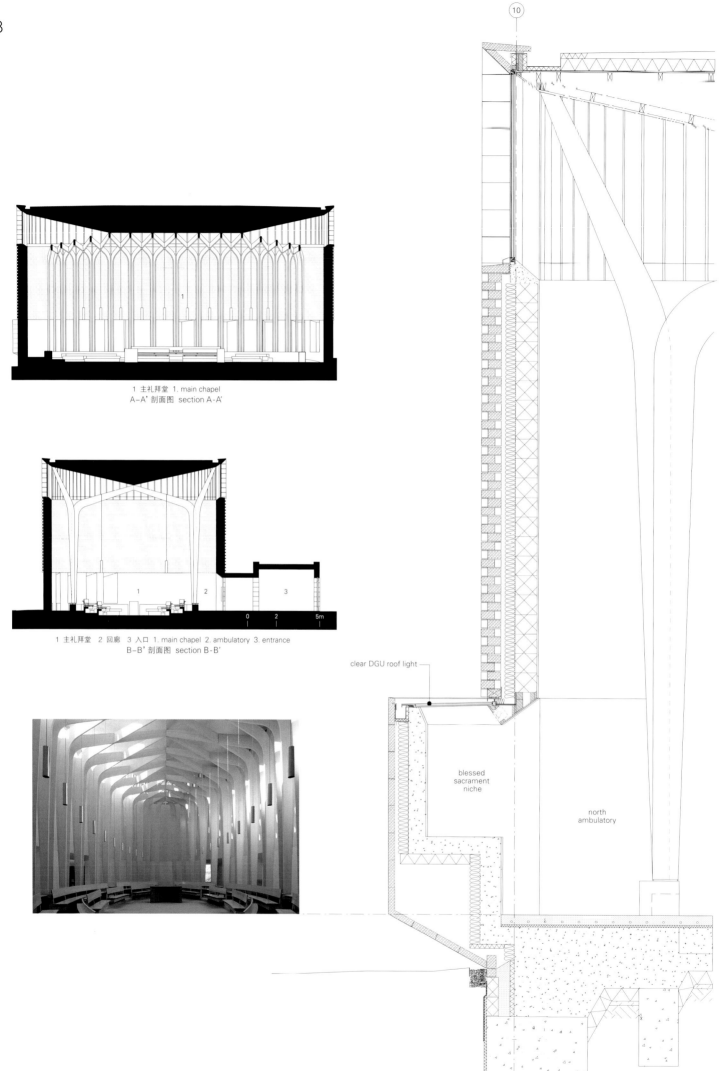

1 主礼拜堂　1. main chapel
A-A' 剖面图　section A-A'

1 主礼拜堂　2 回廊　3 入口　1. main chapel 2. ambulatory 3. entrance
B-B' 剖面图　section B-B'

clear DGU roof light

blessed sacrament niche

north ambulatory

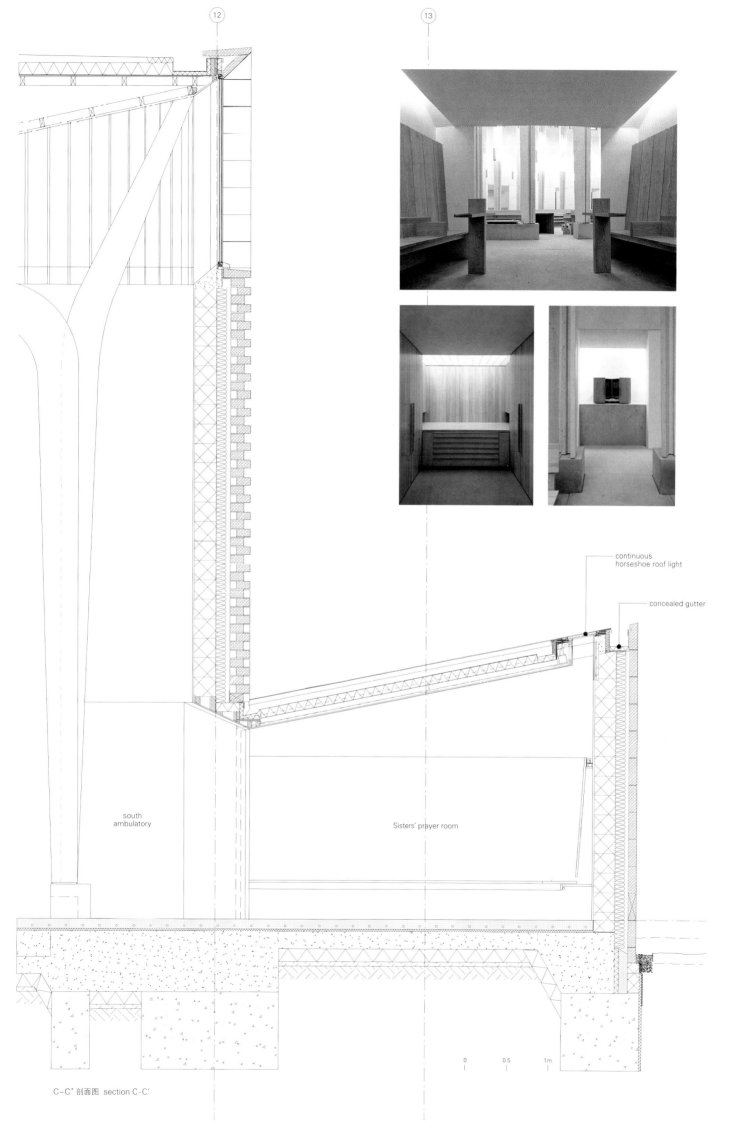

C-C' 剖面图 section C-C'

杰克逊霍尔机场

Gensler

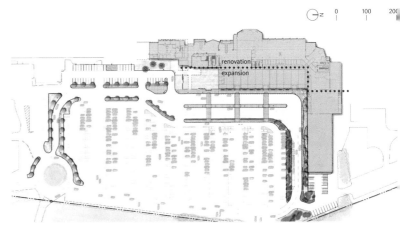

位于怀俄明州的杰克逊霍尔是一个越来越受欢迎的冬夏旅游胜地。它是游览提顿国家公园和黄石国家公园的必经之地，在那里，你可以体验世界级滑雪的乐趣，参加丰富多彩的夏季体育活动。机场往往会让游客留下对杰克逊霍尔的第一印象和最终印象，同时，机场也反映出它位于一个地势相对封闭的社区内，是一个重要的标志，但是发展的空间较小。但重要的是，设计扎根于怀俄明州的杰克逊，反映了其山地环境和西方传统。

杰克逊霍尔机场是美国唯一一座坐落在国家公园的机场。该项目将航站大楼翻新改建，从原来的5521m²扩建到10 738m²。项目涉及扩建售票大厅和停机坪，扩大其面积和体积，改建行李领取处，建造一座全新的、先进的行李安检建筑。这种视觉方面的清晰感与源于整体设计色调来利用绚丽色彩和不同材料，综合而成的寻路策略相互补充，相得益彰。环境责任是杰克逊文化意识的一部分，也是杰克逊霍尔机场的既定目标：最终该项目得到美国第一个LEED绿色建筑银奖认证。

提顿国家公园有严格精确的地界范围划分和5.5m的高度限制，这对项目框架的设计形成了挑战，也对设计团队提出了挑战。传统钢梁或胶合梁的解决方案应该能大大降低天花板的高度，最终将减小其体积。解决方案是要设计一个净跨双柱桁架系统，降低梁高，有效增大体积。此外，由于窗户设计方面的限制，从现有的航站大楼欣赏到的美景非常有限。新扩建的玻璃幕墙将内部和外部有效地连接起来，这样，大量的自然光照射进售票大厅。透明度的增加有助于定位游客方向，产生更加舒适的体验。

大提顿山脉高峻挺拔，气势磅礴，景色壮观，让人震撼。Gensler建筑事务所的理念是建筑纯粹是这美丽风景之中的一个不引人注目的前

景点缀。它存在于景色之中，是令人惊叹的油画布景上的一个元素。该设计并没有将山城的实际情况和历史做参考，而是用现代的方式强调了这一点。这一理念如同在内部和外部之间构建了一个对话框，双方可以展开内容丰富的对话，站在航站大楼便能欣赏到东部和西部广阔的风景。

而在美学方面，杰克逊霍尔机场因其区域的设计方式、材料的使用和亲近自然的程度而有别于其他典型的机场。设计方面可持续发展的属性使杰克逊霍尔机场成为美国少数几个得到LEED认证的机场之一。由于增加了旅游项目，且成为进入该地区的必经之地，杰克逊霍尔机场缓解了航站大楼极端拥堵的状况。

Jackson Hole Airport

Jackson Hole, Wyoming is an increasingly popular tourist destination in both winter and summer. It is the gateway to Teton and Yellowstone National Parks, as well as world-class skiing and a myriad of summer sports and activities. The Airport is a visitor's first and last impression of Jackson Hole and is also an important symbol within a tight-knit local community that is generally averse to growth. It was important that the design is rooted in Jackson, Wyoming, reflecting its mountain environment and western heritage.

Jackson Hole Airport is the only airport in the United States situated in a National Park. The project is a terminal renovation and expansion from 59,426 square feet to 115,578 square feet. The scope of the project includes expansion of the ticketing lobby and hold rooms in both size and volume, renovation of the baggage claim area, and creation of a new state of the art baggage screening

项目名称：Jackson Hole Airport / 地点：Teton County, Wyoming, USA
建筑师：Gensler
合作建筑师：Carney Logan Burke / 结构工程师：Martin Martin
电气工程师：Swanson Rink　土木工程师：Jacobs
行李领取系统设计师：BNP Associates, Inc.
景观建筑师：Hershberger Design
总承包商：Wadman Corporation
用地面积：50,828m² / 总建筑面积：10,740m² / 有效楼层面积：8,547m²
设计时间：2009 / 竣工时间：2011
摄影师：
©Mathew Millman(courtesy of the architect)-p.150~151 p.154, p.157
©Tim Griffith(courtesy of the architect)-p.152, p.153, p.155

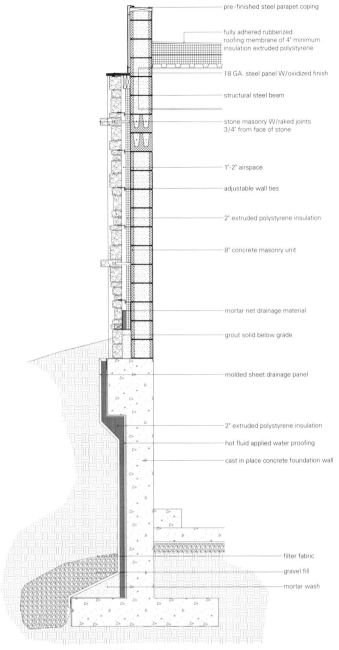

a-a' 剖面图　section a-a'

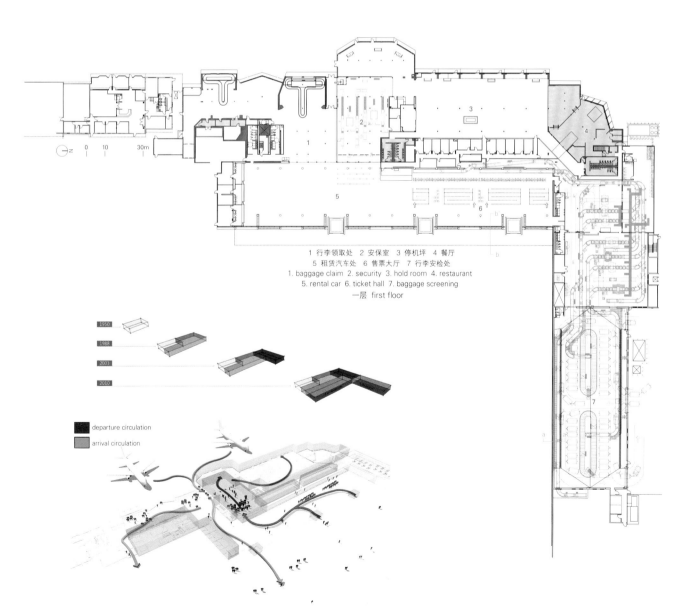

1 行李领取处 2 安保室 3 停机坪 4 餐厅
5 租赁汽车处 6 售票大厅 7 行李安检处
1. baggage claim 2. security 3. hold room 4. restaurant
5. rental car 6. ticket hall 7. baggage screening
一层 first floor

departure circulation
arrival circulation

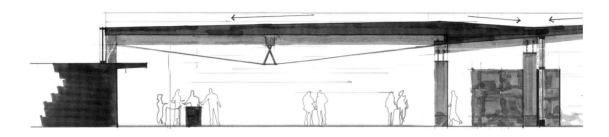

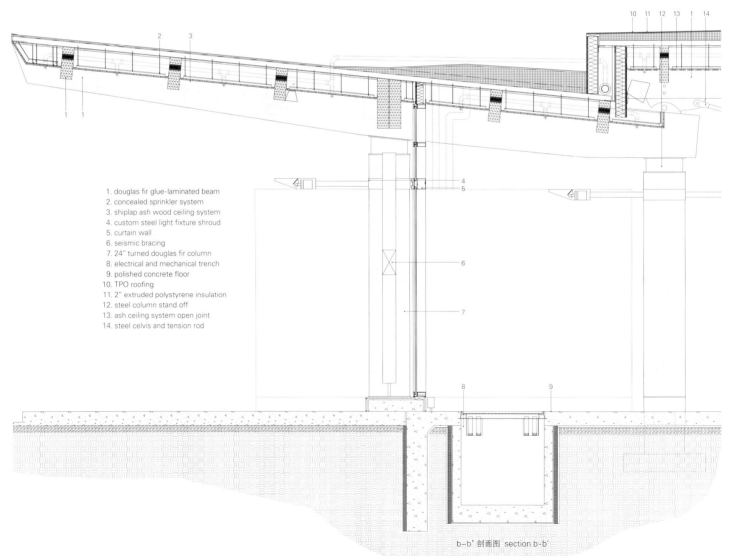

1. douglas fir glue-laminated beam
2. concealed sprinkler system
3. shiplap ash wood ceiling system
4. custom steel light fixture shroud
5. curtain wall
6. seismic bracing
7. 24" turned douglas fir column
8. electrical and mechanical trench
9. polished concrete floor
10. TPO roofing
11. 2" extruded polystyrene insulation
12. steel column stand off
13. ash ceiling system open joint
14. steel celvis and tension rod

b-b' 剖面图 section b-b'

building. This visual clarity is supplemented by an integrated wayfinding strategy that utilizes colors and materials rooted in the overall design palette. Environmental responsibility is part of the cultural awareness of Jackson and the goal of the Airport: as a result the project was the first LEED Silver Certified airport in the country.

Teton National Park has strict site boundary limitations and an 18' height limit establishing a challenging framework for the design. This limitation was a distinct challenge to the design team. A traditional steel or glue lam beam solution would have drastically reduced the ceiling height and consequently the volume. The solution was to design a clear-span queen post truss system that reduced beam depths and effectively increased the volume. In addition, the existing terminal had minimal connection to the wonderful views due to limited fenestration. The new expansive glass curtain wall establishes a strong interior/exterior connection and allows for plenty of natural daylight to flood the ticketing hall. Conversely, the increased transparency helps orient travelers and establishes a more comforting experience.

The exquisite beauty of the landscape and the sheer scale of the Grand Teton Mountain range are overwhelming and humbling. Gensler's concept considered the building as a simple, understated foreground feature within this beautiful landscape. It is intended to merely reside in the landscape, an element within the canvas of the awe-inspiring surroundings. The design addresses this in a contemporary way, not relying on literal and typical mountain town's historic references. The concept establishes a rich dialog between interior and exterior, opening up the terminal to the expansive views to the east and west.

The Jackson Hole Airport distinguishes itself from the aesthetics of typical airports because of its regional design approach, materiality, and intimate scale. The sustainable attributes of the design pave the way for Jackson Hole to be one of only a handful of LEED Certified airports in the United States. The Jackson Hole Airport has alleviated extreme congestion due to increased tourists and becomes a new gateway to the area.

皮塞克市林业局

HAMR

皮塞克市林业局为建造新的办公大楼而进行了招标竞赛,大楼建在森林边上一处美丽的地方,傍依南坡,风景优美。林务员们的小屋就是由走廊从中间一分为二而形成的。它的迷人之处在于大楼周围的所有外部布局——线性的冬季花园,倾斜的遮阳篷——都绕此展开。在这里,建筑与自然的交界面只有几步宽,突破层级式结构的包围,内外相通。建筑的结构本身也呈数层:盒状的经典形式加之支撑雨篷的木质框架结构,其下方的村落内设有一个烘房(带穗的谷物在这里烘干取籽)。建筑师想给这间小屋建个姐妹房,在木材堆叠的形式上完全对称,那里可以作为林业局的信息中心。

Pisek City Forest Administration

Pisek City Forest Administration held a competition for a new operational building on a beautiful site at the edge of the forest; sloping to the south, it has a splendid view. The foresters' lodge is a box sliced in half by the hallway and its charm lies in the way that all of the outer layers of the building are wrapped around it: the linear winter gardens and the slanting shade-awnings. The interface between architecture and nature is a few steps wide, from an interior space surrounded by several layers you find your way to the exterior. The structure also has several layers: the classic form of the box is supplemented with the wooden framing structure that supports the awnings. Down below, at the village, is an oast house (a building where ears of grain are left to dry to release the seeds) and the architect had the idea of making for this little house a kind of sister, a geometrically faithful copy in the form of a lumber stack, and it could be the forest's information center.

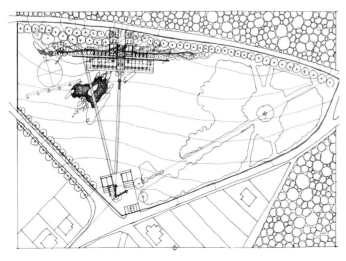

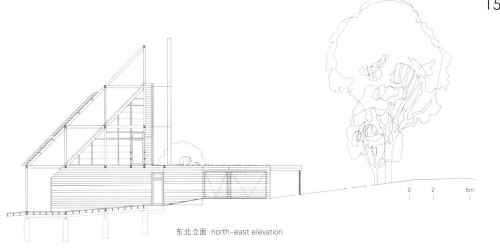

东北立面 north-east elevation

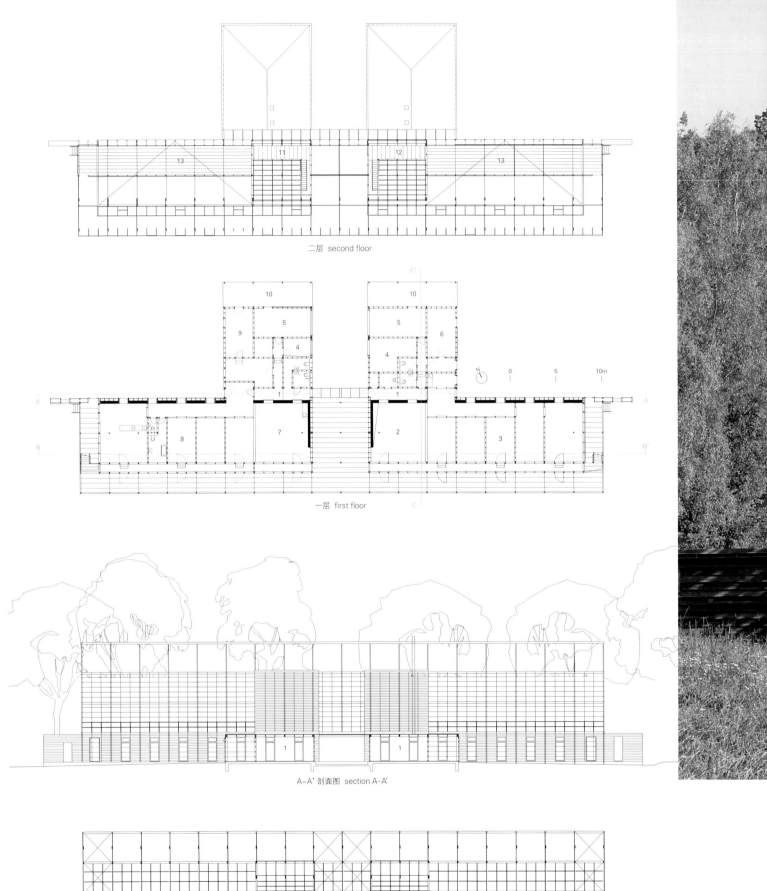

二层 second floor

一层 first floor

A-A' 剖面图 section A-A'

B-B' 剖面图 section B-B'

1	入口	1. entrance
2	会议室	2. conference room
3	办公室	3. office
4	储藏室	4. storage
5	车库	5. garage
6	档案室	6. archive
7	讲室	7. lecture room
8	公寓	8. flat
9	杂物房	9. utility room
10	停车场	10. parking
11	讲室走廊	11. lecture room gallery
12	会议室走廊	12. conference room gallery
13	露台	13. terrace

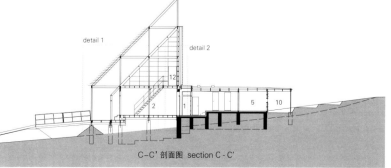

C-C' 剖面图 section C-C'

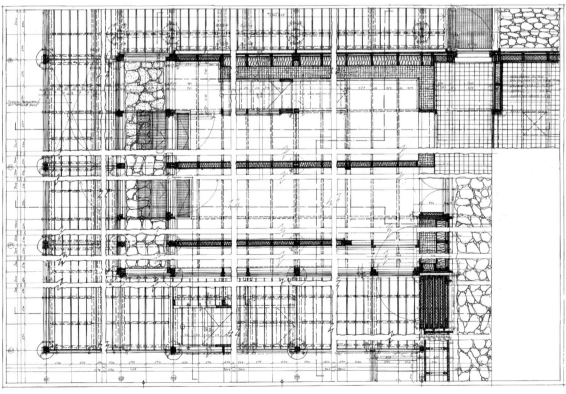

项目名称：Pisek City Forest Administration
地点：Flekačky 2623, Pisek, Czech Republic
建筑师：e-MRAK
项目团队：Martin Rajniš, Martin Kloda, David Kubik
总建筑面积：598m² / 有效楼层面积：520m² / 体积：2,040m³
设计时间：2006 / 竣工时间：2010
摄影师：
Courtesy of the architect - p.162, p.163, p.165 bottom, p.166, p.167, p.168 right, p.169 top
©Radka Ciglerova (courtesy of the architect) - p.158~159, p.161, p.164, p.165 top, p.168 left, p.169 bottom

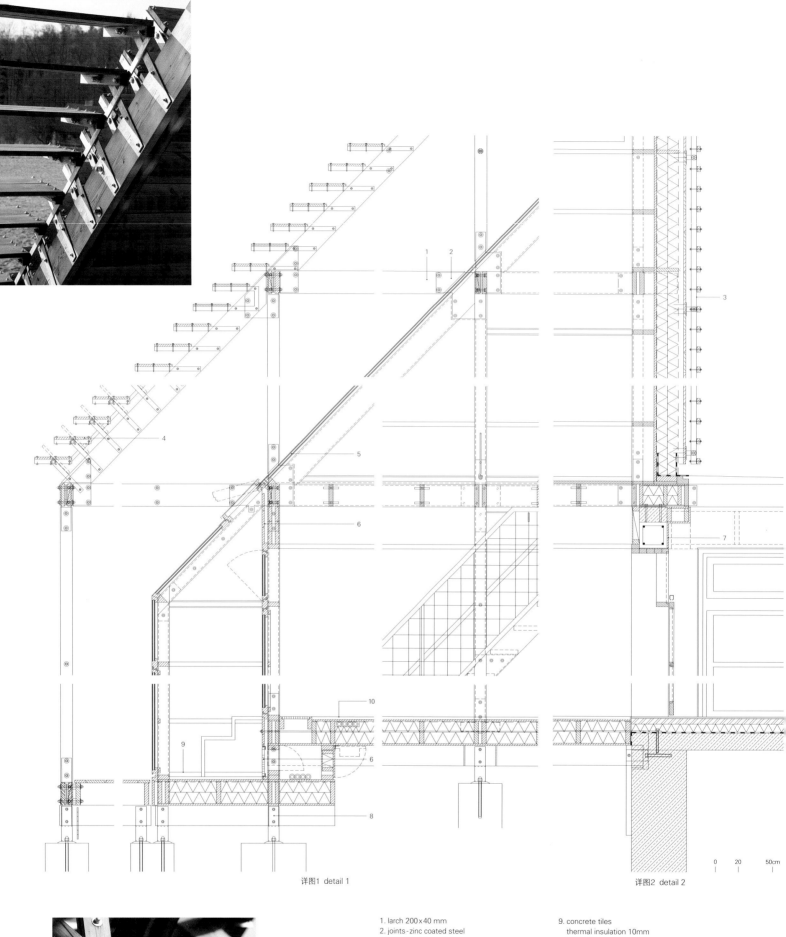

详图1 detail 1   详图2 detail 2

1. larch 200×40 mm
2. joints - zinc coated steel
3. larch 30×50 mm/air
   larch planks 20mm/air 50mm
   load bearing construction-larch 40×220mm
   mineral wool 220m
   spruce planks 20mm
   vapour barrier
   load bearing construction-spruce 40×220mm
4. adjustable shadowing system
5. insulating glass in wooden frames
6. ventilation
7. masonry, plaster
8. anchoring - zinc coated steel
9. concrete tiles
   thermal insulation 10mm
   OSB 15mm
   vapour barrier
   load bearing construction-larch 40×220mm
   mineral wool 220m
   OSB 15mm
10. tal oil/spruce planks
    acoustic pad 3mm
    OSB 15mm
    vapour barrier
    load bearing construction-larch 40×220mm
    mineral wool 220m
    OSB 15mm

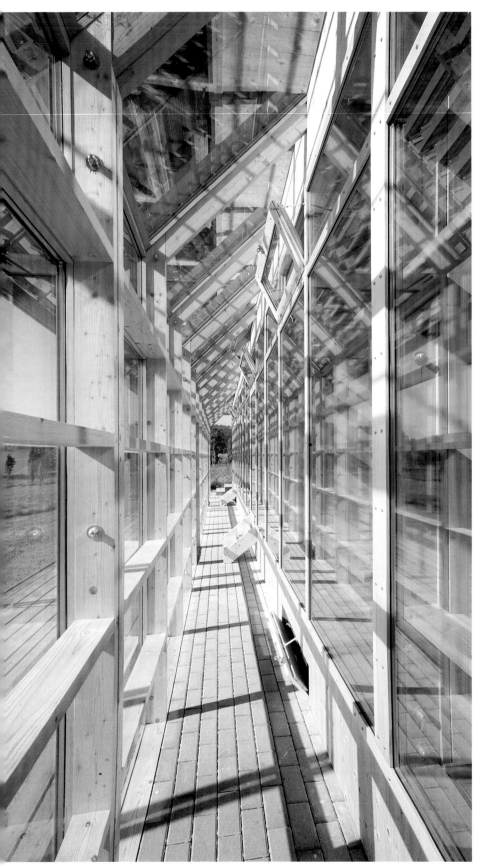

园艺展览中的展亭

Dethier Architecture

2011年，科布伦茨市被选定为德国联邦园艺博览会举办地，这是一个两年一次的博览会，吸引了360万游客。当地为这一盛会新建了各式建筑，包括Dethier建筑事务所设计的观景楼。除了观景以外，该建筑也已成为这一城市的象征。

德国联邦园艺博览会是规模宏大的盛会，科布伦茨市需要能够迎接这一挑战的基础设施。准备工作的一部分就包括观景楼的国际招标。观景楼拟建在艾伦布来斯坦要塞边上，可以俯瞰到摩泽尔山谷和莱茵河的交汇处。这一要塞在一战后成为人们喜爱的重要的旅游景点，也荣登联合国教科文组织的世界遗产名录。

观景楼呈一个中空的三角形，坐落于高原之上，俯瞰着科布伦茨市。游客可以找到一条便捷的走廊，走廊从用于展览的美术馆一直延伸到屋顶，沿廊而行，公园、城市、观景楼本身尽在眼底。悬臂结构是这一项目的亮点：它从山谷向外延伸超过15m，离开地面10m。结构和走廊是采用当地的树木制成的，结构构件的材质则为考顿钢，这些材料的选择意味着整个结构可以在工厂预制。建筑研究和工程的结合确保了稳定性，其结果是轻质的观景楼设有动态的游客环线。侧面的桁架结构点点映衬着周围的乡村景色，并使得观景楼可以相对独立地存在。

以其正规、高科技的设计，观景楼展现了现代化的视野，最大限度地将环境特征交融在一起，反映了体察自然的意愿。建筑师需要找到一个正确的解决方案以便以一种完全朴素的风格，使这一建筑能够成为这一城市的象征。此外，能够让每一个游客与周围的环境切实且微妙地接触也至关重要。基于当地地貌考察的结果，对花旗松板材以及其他材料的选择都满足了上述考虑。如此规模宏大的建筑，很远就能看见，它的魅力与横向延展使其与山谷的风景和谐交融。周围公园的特点也是决定性因素。设计者请求这一建筑尊重当地的历史性，对此，观景楼的位置和三角形设计都可以反映要塞的基础，花园小径似延伸出谷的悬臂，德国之角就在其下。

Pavilion for Horticultural Show

In 2011, Koblenz was chosen as the site for the Bundesgartenschau, a biennial horticultural show that drew 3.6 million visitors. Various structures were erected for the occasion, including a belvedere designed by Dethier Architecture. In addition to providing a lookout point, the belvedere has become a symbol of the city.

The Bundesgartenschau is a large-scale event, and the city of Koblenz needed an infrastructure capable of meeting the challenge. Part of the preparations involved an international competition for the construction of a belvedere, to be built on the site of the Ehrenbreitstein Fortress, which overlooks the confluence of the Moselle and Rhine rivers. The fortress is an outstanding site that became a popular tourist attraction after the First World War, and it has been placed on UNESCO's World Heritage List.

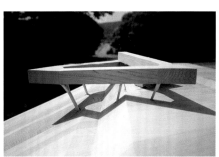

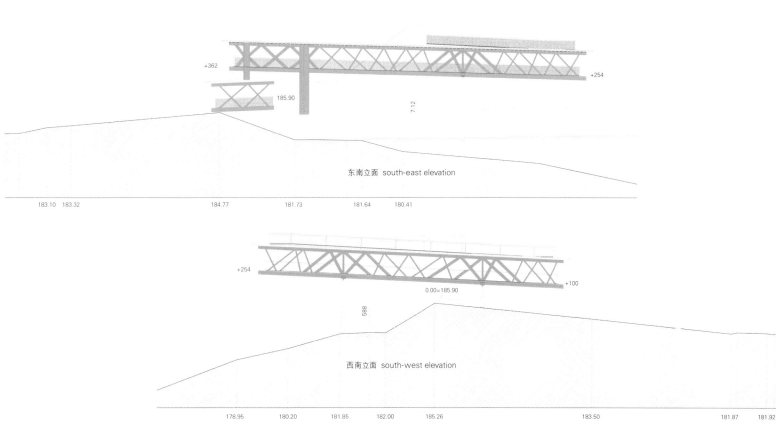

The belvedere is in the shape of a hollow triangle positioned on the plateau overlooking Koblenz. A walkway, accessible to visitors with limited mobility, leads from a gallery – a potential exhibition space – to the roof, along a pathway offering, by turns, views of the park, the city and the belvedere itself. The cantilever symbolises the project: it extends more than 15 metres out over the valley, and rises 10 metres above the ground. The choice of materials (native wood species for the structure and walkway, and corten steel for the structural elements), meant that the entire construction could be pre-fabricated. The marriage of architectural research, and engineering to ensure stability, has resulted in a lightweight structure with dynamic visitor circulation. The lateral trusses create a mosaic of the surrounding countryside and allow the structure to be relatively free-standing.

With its formal, high-tech design, the belvedere offers a vision of modernity. The integration of a maximum number of contextual characteristics reflects a willingness to foreground the perception of nature. A correct solution needed to be found so that, in a completely unostentatious manner, the structure could stand as a symbol of Koblenz. Moreover, it was essential to enable each visitor to have a meaningful, subtle encounter with the surroundings. The choice of Douglas fir and other materials – and above all an examination of the site's morphology – allowed this "architectural object" to encompass these concerns. As a large-scale construction, the structure can be seen from afar, but its grace and horizontality allow it to blend harmoniously into the valley's landscape. The surrounding park's features were also determining factors. In response to the organisers' request that the project respects the historic nature of the site, the belvedere's location and angular design were chosen to reflect the fortress' foundations and the garden pathways, which appear to cantilever out over the valley and the Deutsches Eck.

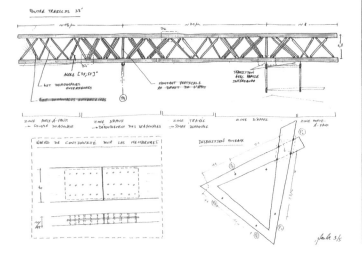

项目名称：Belvedere for the 2011 Bundesgartenschau
地点：Bleidenberg, 56068 Koblenz, Germany
建筑师：Dethier Architecture
结构工程师：Ney & Partners
总承包商：Mohr Ingenieurholzbau GmbH
甲方：Bundesgartenschau Koblenz 2011 GmbH, Landesforten Rheinland-Pfalz
表面积：675m²
获奖情况：Anerkennung, Holzbaupreis Rheinland Pfalz 2011, Skulpturales Bauen
竣工时间：2011.5
摄影师：©Thomas Faes & Polizeiprasidium (courtesy of the architect)

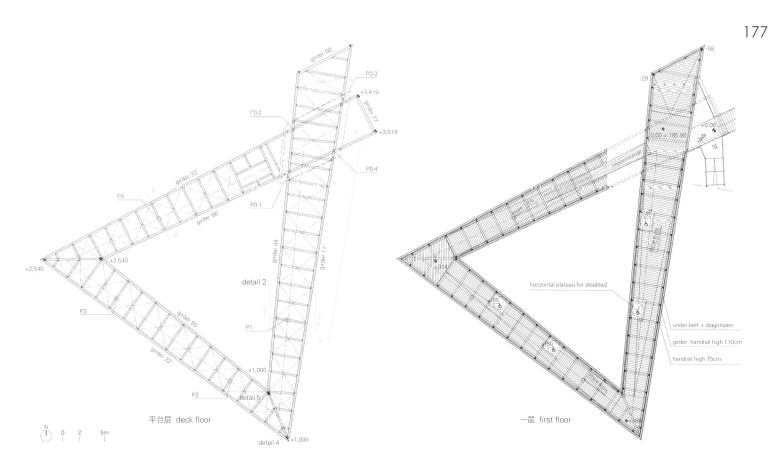

平台层 deck floor

一层 first floor

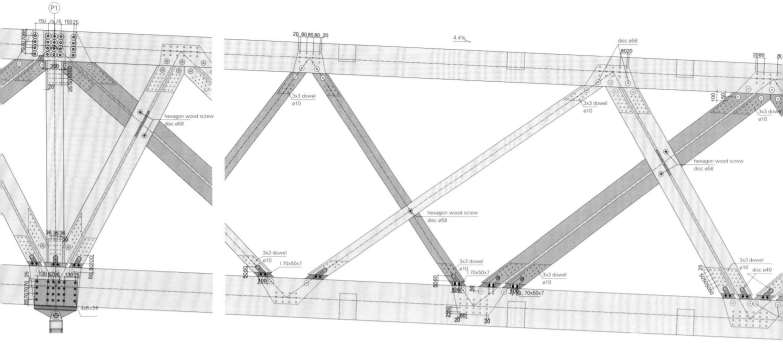

a-a' 立面图_主梁
elevation a-a'_ girder 11'

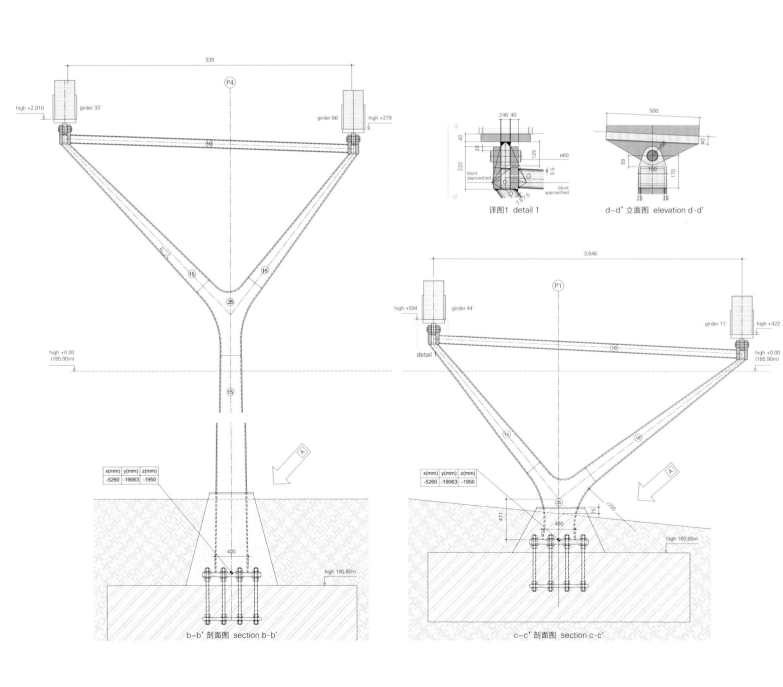

详图1 detail 1

d-d' 立面图 elevation d-d'

b-b' 剖面图 section b-b'

c-c' 剖面图 section c-c'

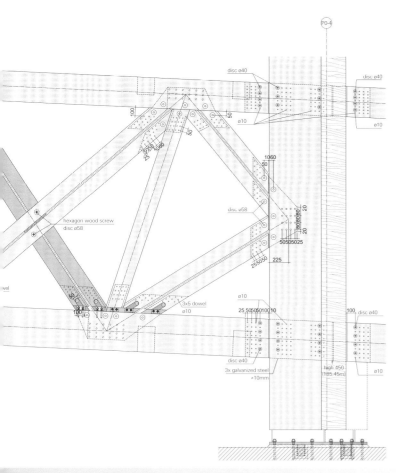

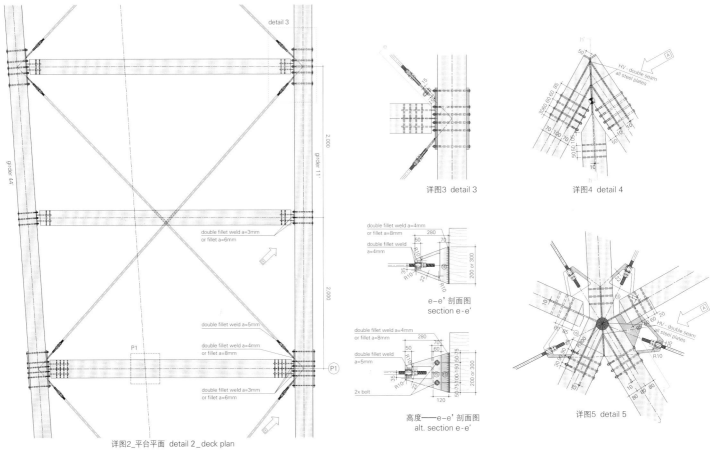

详图2_平台平面 detail 2_deck plan

详图3 detail 3

详图4 detail 4

e-e' 剖面图 section e-e'

高度——e-e' 剖面图 alt. section e-e'

详图5 detail 5

详图6 detail 6

f-f' 剖面图 section f-f'

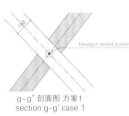

g-g' 剖面图 方案1
section g-g' case 1

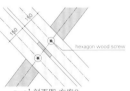

g-g' 剖面图 方案2
section g-g' case 2

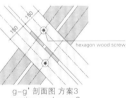

g-g' 剖面图 方案3
section g-g' case 3

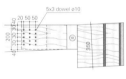

h-h' 剖面图 section h-h'

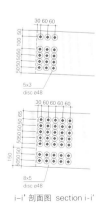

i-i' 剖面图 section i-i'

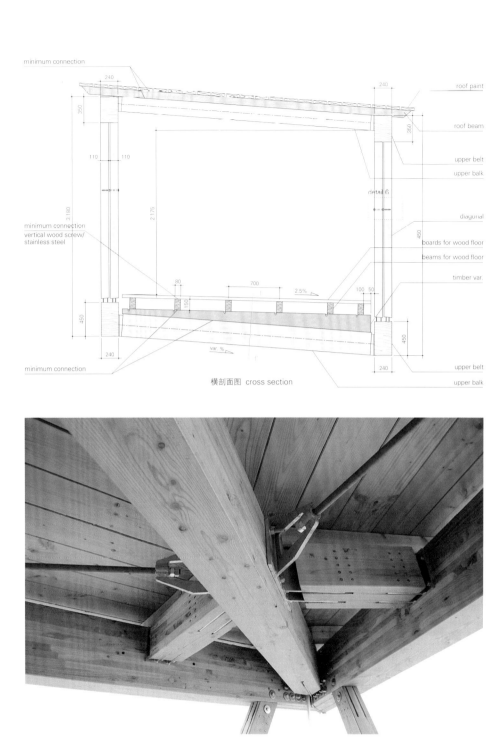

横剖面图 cross section

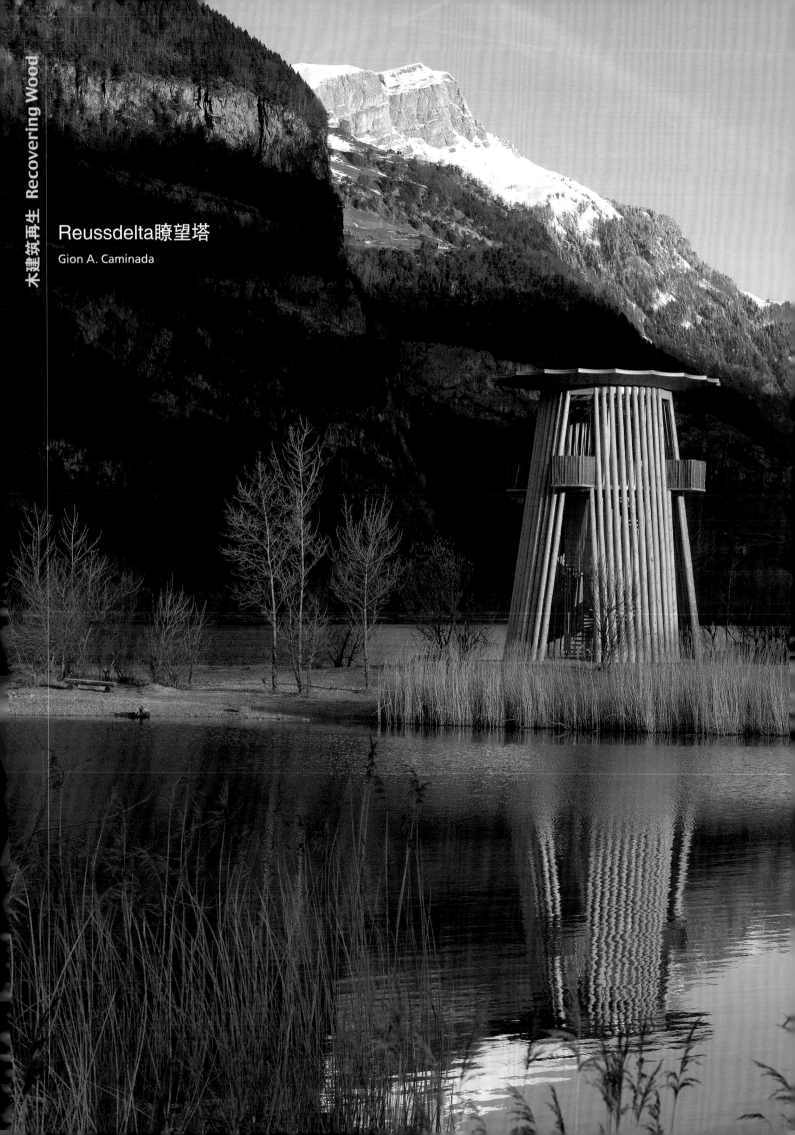

木建筑再生 Recovering Wood

Reussdelta瞭望塔
Gion A. Caminada

大量的动物和植物生活在琉森湖沿岸。瞭望塔的建造权衡了对自然的保护和使用。然而，这座塔不仅仅作为瞭望地点。编筐工人已经把这座塔变成了工艺品。工艺和材料的紧密关系增进了这座塔在本地的识别性。

Reussdelta地区是一处文化景观。文明化使此地成为疗养和逃离日常生活的地方。然而，准确地讲，这个文明化也是一个冒险。这种分裂形成的紧张关系是很难描述的。这座塔不是以当地的建筑传统为基础，而是以施工过程中的建筑原型的形式为基础，避免成为占有性的地标建筑。

此塔没有倾向的朝向。因此，人们可以由四个方位进入其中。开口处也不是简单的入口，而是融合到结构形成的结果中。行走于旋转楼梯上的人们能够到达带有四个阳台的观察平台上。人们既可以露面也可以隐藏在被屋顶保护的平台中心中。同样，屋顶覆盖在木质结构之上，提供了保护作用。楼梯和阳台的栏杆由柳条编制而成。光从缝隙中透过来，形成了空间氛围。塔建在混凝土板上。混凝土中使用了河卵石。原始云杉木材向内倾斜，进而形成圆柱。这个结构的中心由木质圆柱组成，设有楼梯和屋顶。承重、支撑和悬浮构件之间的相互作用形成了空间网络。一个构件决定了其他构件，最终形成了木质建筑的连贯性。

Reussdelta Observation Tower

Numerous animals and plants reside along the shore of lake Urnersee. The tower was built within the trade-off between preservation and usage of nature. However the tower serves not only as a place for observation. The basket maker's work has made the tower a place of craftsmanship. The close relationship to the processes and materials involved gives rise to the growth of a local identity.

The area of Reussdelta is a cultured landscape. Civilization formed the area as a place of retreat and escape from everyday life. However precisely this civilization is also one peril. This dichotomy creates tension which one can hardly describe. The tower is not based on regional architectural traditions. Instead the tower is rather based on archetypal forms of construction. The tower avoids being a possessive landmark.

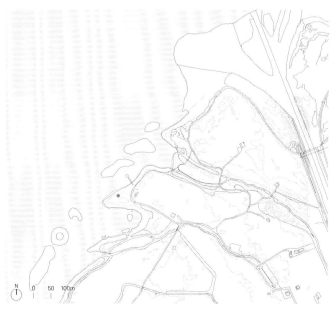

The tower does not want to have a preferred orientation. To this end the tower is accessible from four directions. The openings do not want to be simple entrances, they emerge as a consequence from the structure. Walking on the spiral stairs one reaches the viewing platform with its four balconies. One may either show up or stay hidden in the center of the platform, protected by the roof. Likewise the roof acts as a protecting shield for the timber construction. The railings of stairs and balconies are woven from willow timber. Light can enter through gaps and set the spacial atmosphere. The tower is built on a concrete slab. River gravel is used for the concrete. The raw spruce logs lean inward and thereby form a cylinder. The center of the structure is formed by a wooden column. It also contains the stairs and carries the rooftop. The interplay between load-carrying, supporting and suspending elements creates a spacial network. One element determines the other which results in a coherent piece of timber architecture.

项目名称：Aussichtsturm Reussdelta
地点：Seedorf, Switzerland
建筑师：Gion A. Caminada
工程助理：Jan Berni
施工工程师：Pirmin Jung
甲方：Canton Uri
竣工时间：2013
摄影师：©Lucia Degonda (courtesy of the architect)

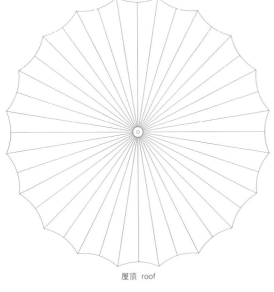

屋顶 roof

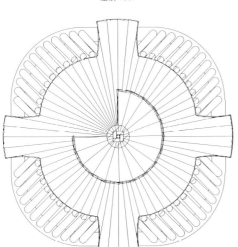

二层 second floor

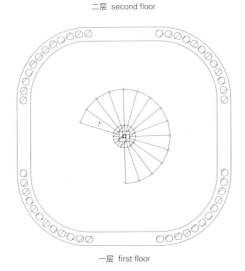

一层 first floor

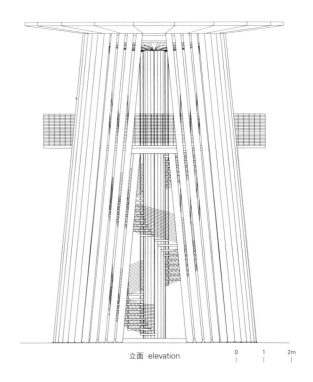

立面 elevation

>>170
Dethier Architecture
Daniel Dethier studied architecture and civil engineering at the university and graduated in 1979. He founded Dethier Architecture in 1992. Free of all prejudice, the procedures are characterized by a critical vision allowing them to provide innovative solutions appropriate to current requirements in fields as varied as the type of habitat in towns and in the country, adding value to our heritage, cultural infrastructures, development of public areas, urban planning, design of interiors, furniture or equipment.

>>128
Shigeru Ban Architects
Shigeru Ban was born in Tokyo, 1957. Excelled at arts and crafts throughout school years then decided to become an architect. Received his Bachelor of Architecture from Cooper Union in 1984 and started his own practice in Tokyo without any work experience in 1985. Through various working experiences after that, he became a professor on the Faculty of Environment and Information Studies at Keio University in 2001 and worked until 2008. After he won the competition of Centre Pompidou-Metz, he established a private practice in Paris with his partner Jean de Gastines. Ban is currently working on creating architecture, he volunteers for disaster relief, lectures widely, and teaches. He continues to develop material and structure systems.

>>138
Níall McLaughlin Architects
Níall McLaughlin was born in Geneva in 1962 and educated in Dublin. Received his architectural qualifications from University College Dublin in 1984. Established his own practice in London in 1990. He was chair of the RIBA Awards Group from 2007 to 2009. Was also a visiting professor of architecture at University College London as well as a visiting professor at the University of California Los Angeles from 2012 to 2013 and appointed as Lord Norman Foster Visiting Professor of Architecture, Yale for 2014~2015. He lives in London with his wife Mary, son Diarmaid and daughter Iseult.

>>96
Fraher Architects
Was established in 2009 by Joe Fraher[left] and Lizzie Webster[right]. They work primarily in the private and commercial residential sectors with a high planning success rate for contemporary developments within sensitive sites.
Deliver creative, yet accessible designs, always drawing from a project's unique relationship to its history and context anchoring it into the built and natural environment. Their design strategy assesses each project as a unique architectural response, introducing experiential spaces as opposed to spaces based on theories. Through this approach, they believe that space, whether private or public, can be individual, dynamic and beautiful. Aim to design spaces that are not consumed by the ideals of the architect, but to deliver projects designed and built by a team for a common purpose.

>>150
Gensler
Brent Mather, a senior associate (regional design leader), has joined Gensler in 1999. Received Bachelor in Architecture from University of Tennessee. Won the AIA Colorado Young Architect of the Year Award in 2009. Is a member of AIA and the U.S. Green Building Council (USGBC). Is also a terminal planner and regional design director. Has led the design of the firm's award-winning projects of Jackson Hole Airport, Tulsa International Airport Renovation, and Will Rogers World Airport – all which began as Terminal Planning projects. His passion is searching for the inherent harmony that exists between architecture, site, and the people who experience it.

>>102
Carlo Bagliani
Was born in Genova, Italy in 1965. Was a partner of an architectural studio Sp10 until 2011 and opened his own office in 2012.

©Manfred Klimek

>>54
Coop Himmelb(l)au
Was founded in 1968 by Wolf D. Prix, Helmut Swiczinsky and Michael Holzer in Vienna, Austria. Is active in architecture, urban planning, design, and art. In 1988, a second studio was opened in Los Angeles, USA and further project offices are located in Frankfurt, Germany and Paris, France. Recently, won the MIPIM Architectural Review Future Projects Award in the sustainability category for the Town Erdberg and received the Wallpaper Design Award 2011 in the Best Building Sites category.

>>68
Yuko Nagayama & Associates
Yuko Nagayama was born in Tokyo in 1975. Graduated from Showa Women's University in 1998 and established Yuko Nagayama & Associates in 2002 after working at Jun Aoki & Associates. Received numerous awards including Architectural Record Award, Design Vanguard 2012. She is teaching at Kyoto Seika University, Showa Women's University, Ochanomizu University, and Nagoya Institute of Technology. Representative work includes Louis Vuitton Kyoto Daimaru, Anteprima Singapore, Kiya Ryokan etc. Continues to participate on various exhibitions like Kenchiku Architecture and solo exhibition "Exhibition of Yuko Nagayama".

>>158
HAMR
Martin Rajniš was born in 1944, Praha. He has studied at the Faculty of Civil Engineering and Architecture School of Fine Arts in Praha. Founded HAMR in 2012. Has been sailing and traveling most of the European and American countries for over 50 years. Sailing and journeys mean a lot for him as well as his houses and books.

>>182
Gion A. Caminada
Was born 1957 in Vrin, Switzerland. Following a carpenter apprenticeship in Vrin, he attended the Kunstgewerbeschule in Zurich and then went on to complete postgraduate studies in architecture at the ETH Zurich. Since 1986, Caminada has had his own architectural firm in Vrin. Within the span defined by teaching, research and practice, he has been dealing for many years with building culture in a comprehensive sense. In 1999, he was named as an assistant professor at ETH Zurich, and since 2008 he has been an associate professor for architecture.

Silvio Carta
Is an architect and researcher based in London. Received Ph.D. from University of Cagliari, Italy in 2010. His main fields of interest is architectural design and design theory. His studies have focused on the understanding of the contemporary architecture and the analysis of the design process. He taught at the University of Cagliari, Willem de Kooning Academy of Rotterdam and Delft University of Technology. He is now a senior lecturer at the University of Hertfordshire. Since 2008 he is editor-at-large for C3, Korea and his articles have also appeared in A10, Mark, Frame and so on.

>>40

Morphosis Architects
Is an interdisciplinary practice involved in rigorous design and research that yields innovative, iconic buildings and urban environments, founded in 1972. With founder Thom Mayne serving as design director, the firm today consists of a group of more than 40 professionals, who remain committed to the practice of architecture as a collaborative enterprise. With projects worldwide, the firm's work ranges in scale from residential, institutional, and civic buildings to large urban planning projects. Over the past 30 years, Morphosis has received 25 Progressive Architecture awards, over 100 American Institute of Architects (AIA) awards, and numerous other honors.

Douglas Murphy
Studied architecture at the Glasgow School of Art and the Royal College of Art, completing his studies in 2008. As a critic and historian, he is the author of The Architecture of Failure(Zero Books, 2009), on the legacy of 19th century iron and glass architecture, and the forthcoming Last Futures (Verso, 2015), on dreams of technology and nature in the 1960s and 1970s. Is also an architecture correspondent for Icon Magazine, and writes regularly for a wide range of publications on architecture and culture.

>>110

Akitoshi Ukai / AUAU
Was founded in 2000 by Akitoshi Ukai. He graduated from Aichi Institue of Technology and completed

Tom Van Malderen
His activities stretch from the traditional architectural practice to the field of architectural theory which he explores through writing, installations and lectures. After obtaining a master in Architecture at LUCA, Brussels(1997) he worked for Atelier Lucien Kroll in Belgium and in different positions at architecture project, both in the UK and Malta. Lectured at the University of Aix-en-Province in France and the Canterbury University College of Creative Arts in the UK. Contributes to several magazines and publications, and sits on the board of the NGO Kinemastik for the promotion of short film.

>>118
Savioz Fabrizzi Architectes
Founded in 2004 by the two architects Laurent Savioz[right] and Claude Fabrizzi[left], is trying to respond with the best conditions to the needs of the clients by providing all the architectural services from the project to the achievement. Their work is based on the analysis of a site in its natural or built state in order to identify the essential elements that could enhance, preserve or qualify a site. In this way the firm enhances the cultural role of the architecture based on the analysis of a function, respectively a program, its place in the history and the culture of a region.

>>74
Haerynck Vanmeirhaeghe Architecten
Tijl Vanmeirhaeghe is a Belgian architect-engineer(UGent, 1999) and teacher. Was born in Ghent, 1976 and studied Civil Engineer Architect at the University of Ghent, 1994-1999. Was awarded the Young Belgian Designer Award in 2000 and shortly afterwards founded BARAK architects (2004-2011) together with Carl Bourgeois. Their work, mainly housing projects and scenographies, was published in several books and magazines and exhibited on different occasions. Since 2012 he's running the new office Haerynck Vanmeirhaeghe Architecten together with Bert Haerynck.
Tijl Vanmeirhaeghe is an associate professor in architectural sciences and design, at Ghent University, Department of Architecture and Urban Planning since October 2008. He published several articles and contributed to various publications on art and architecture by the department.

>>84
Estudio Elgue
Luis Alberto Elgue Sandoval is a chief architect of Estudio Elgue and a partner based in Asuncion, Paraguay. Graduated from the National University of Asuncion(UNA). Has taught courses and seminars in São Paulo(postgraduate course 2011) Guatemala(2012), Colombia(2013), Cuba (2014). Is now a board member of the Faculty of Architecture at the National University of Asuncion.

C3, Issue 2014.6

All Rights Reserved. Authorized translation from the Korean-English language edition published by C3 Publishing Co., Seoul.

© 2014 大连理工大学出版社
著作权合同登记06-2014年第124号

版权所有·侵权必究

图书在版编目(CIP)数据

木建筑再生：汉英对照 / 韩国C3出版公社编；杨惠馨等译. —大连：大连理工大学出版社，2014.8
(C3建筑立场系列丛书)
书名原文：C3 Recovering Wood
ISBN 978-7-5611-9366-2

Ⅰ. ①木… Ⅱ. ①韩… ②杨… Ⅲ. ①木结构－建筑设计－汉、英 Ⅳ. ①TU366.2

中国版本图书馆CIP数据核字(2014)第164909号

出版发行：大连理工大学出版社
　　　　　（地址：大连市软件园路80号　邮编：116023）
印　　刷：上海锦良印刷厂
幅面尺寸：225mm×300mm
印　　张：12
出版时间：2014年8月第1版
印刷时间：2014年8月第1次印刷
出 版 人：金英伟
统　　筹：房　磊
责任编辑：张昕焱
封面设计：王志峰
责任校对：赵姗姗

书　　号：978-7-5611-9366-2
定　　价：228.00元

发　行：0411-84708842
传　真：0411-84701466
E-mail：dutp@dutp.cn
URL：http://www.dutp.cn